Call of the Kyeema

Doug Whitfield

ISBN 0 9751289 4 9

All rights reserved. No part of this publication may be reproduced, stored in a retrieval system, or transmitted in any form, or by any means - electronic, mechanical, by photocopying, scanning, recording or in any other way without the prior agreement of the publisher in writing.

Printed and Bound by Success Print
61-8-9279 3214
7a Goongarrie Street
Bayswater
Western Australia 6053

Copyright © 2002 Doug Whitfield

Preface

Doug Whitfield joined the Department of Civil Aviation in 1969 as a Flight Service Officer and worked at remote locations in South Australia, Western Australia and the Northern Territory. Following this service he became an Air Traffic Controller in Victoria, Queensland, Western Australia and the Northern Territory . He served as an officer in both the Royal Australian Air Force and the Australian Army and flew as a Commercial pilot in Queensland and Western Australia. He was the Town Council's Executive Manager for Port Hedland International Airport and later moved back to the Northern Territory where he became the Manager at Ayers Rock (Connellan) Airport. He then moved to Darwin and worked as the Operations Manager at Darwin International Airport before accepting a position with the Civil Aviation Safety Authority as the Area Manager for the Northern Territory and Kimberly Region of Western Australian. He is still heavily involved in Australian aviation and has written articles for various magazines.

The twenty first century has arrived with its new technology. Advances in radio and satellite communications, computer controlled information services and microwave data links have not paid heed to the doomed cry of the Kyeema in the Australian Dandenong Ranges many years ago. The loss of its crew and passengers has been lost to time, but the crash is remembered by many who worked in aviation and watched the spectacle of aviation controls unfold. Sadly, the Aeradio system that evolved from that fateful day has been overtaken by time, and its descendant Flight Service and the Operational Control method of flight checking has also vanished. Doug now recalls the period from 1968 that followed his journey around Australia .This is not a historical anthology but a book of stories that gives a unique personal insight into an era and the personalities that have almost been forgotten .The yarns that come from this trip down memory lane will be enjoyed by everyone, whether they were employed in aviation or not.

By the time the creek had reached the end of the mountains, it had matured into a river. And being mature, it no longer splashed and played like it did at the beginning, but now moved with the certainty of its direction assured of its ultimate destiny into the sea.

Anon

To my wife Katherine.

Mr W.S Whitfield,

Mr E.H. and Glenys Whitfield

my sons Scott and Steve

my daughter Trish

Mr C.B. Hulse OIC DCA Flight Service School

and

especially to the men and women

who dedicated themselves

to the Aeradio/Flight Service Units

around Australasia from 1950 until 1990

Acknowledgments

I am indebted to Mr C.B.(Syd) Hulse who assisted with several items of information provided from interstate by telephone in 1990. I would also like to thank the following people who assisted me with my research and ideas. Major Gordon Issom from the Australian Army School of Signals at Watsonia, Mr Clive Cassin and Mr Reg Vagg from the Civil Aviation Authority in Perth, Maurice Lowe from SCEL Pty Ltd at Essendon airport, Mr Roger Meyer from the Civil Aviation Historical Group in Melbourne for his invaluable assistance, Ken Munro in Brisbane, Phil Parker, now resident in Hong Kong, Harry Gentle from Queensland, Lindsay Telford from Victoria, Andrew McMillan from Darwin, Phil McCullough, Tom Saggers and Jeff Dittrich. My grateful thanks to my wife Kathy for her support in listening to me reading the changes to the book sections every day of her life and for her amazing patience, calm and encouragement during the writing of the manuscript. Thank you all.

Doug Whitfield.

Publishers Dedication

The publishing of this work has been both a privilege and a pleasure. Many events and reminiscences herin have recalled to my mind my late friend John Bell, with whom I share many memories of light aircraft, whale spotting and flying in the Albany region. To John's memory I dedicate my contribution and efforts.

Tom Saggers
November 2004

Contents

	List of Photos and Figures	vii
1.	Cry from the Kyeema	1
2.	The Fledgling	5
3.	Early DCA Days	12
4.	First Bush Posting	21
5.	Darwin Duties	36
6.	Gove on the Gulf	62
7.	A Tale from the Track	78
8.	Alice Springs Adventures	86
9.	Freds Flight	99
10.	Tiger Moth Tale	105
11.	Charlie Mike India	112
12.	Oh You Fly?	118
13.	Search	124
14.	Spencer Gulf	134
15.	Captain Arthur Turner	145
16.	Wings in the West	165
17.	North Queensland	183
18.	My moment of Destiny	204
19.	Post Crash Northings	208
20.	Letter from New Guinea	222
21.	Aerodromes	251
22.	The Winds of Change	264
~	Glossary	272
~	In Memorium	276
~	Flight Service & Comm Officers 1969-1976	277

List of Photos and Figures

1. Photo – The author reads the plaque on the long forgotten memorial cairn built 50 metres above the Kyeema crash site on Mt Dandenong Victoria (Author)   3
2. Memorial plaque at Canberra (Author)   4
3. Victa Airtourer V115 two seater trainer (Author)   7
4. The DCA Visual Flight Guide from the 1960s era (Author)   9
5. DCA Ford Falcon 1970 (DCA)   11
6. DCA Central Training College Henty House Melbourne 1970 (Author)   12
7. First Aeradio Officers Course 1940 (Civil Aviation Historical Society)   15
8. Typical permanent Aeradio room (Civil Aviation Historical Society)   16
9. Flight Service course no 11 graduated Melbourne July 1970 (Author)   20
10. Flight Service Don Hodder operating the Katherine VHF (Author)   26
11. Northern Territory Aerial Services Medical De-Havilland Dove (Author)   30
12. The disused Katherine Flight Service Unit 1997 (Author)   30
13. Connellan Airways Heron DH114 registered VH-CLT (Author)   35
14. Mr Lindsay Telford on duty in the Darwin Aeradio Station 1950 (L.Telford)   36
15. One of the disused RAAF huts Parap 1970 (Author)   40
16. The Darwin Flight Information Region Operations Building 1997 (Author)   41
17. Fight information sector strip on the Singapore Darwin route (Author)   47
18. DCA Communications Officer Terry Ford 1970 (Author)   50

19. Old Alice Springs Briefing Office with Ansett FK27 crew (Author)	52
20. Darwin Aero Club 1970 (Author)	57
21. After my first solo on 8 December 1970 (Author)	57
22. Darwin city taken from Victa 115 at 1500 feet (Author)	58
23. Historic telegram details re bombing of Darwin (Author)	59
24. Map of wartime airstrips at Darwin (Author)	61
25. Headstone of Pilot Officer Gove (Author)	63
26. The vacated EUROPA site at Gove NT (Author)	67
27. Author outside the Gove Flight Service Unit 1970 (Author)	68
28. Gove Flight Service Unit as a passenger terminal 2001 (Author)	68
29. Gove Flight Service Unit Communications Consolette 1971 (Author)	69
30. Gove Emergency runway lighting flare trailer in 1971 (Author)	72
31. Wreckage of an RAAF Ventura bomber at Gove 1971 (Author)	75
32. Remains of an RAAF Bristol Beaufighter Gove 1971 (Author)	76
33. Stuart Highway flood (Author)	83
34. Aerial shot of the Alice Air Terminal 1970 (Author)	86
35. John Scougall at the old Alice Springs radio console 1972 (Author)	87
36. Alice Springs floods 1972 at Heavitree Gap (Author)	91
37. Doug Whitfield at the new Alice Springs radio console 1972 (Author)	92
38. Unexpected international arrivals at Alice Springs 1972 (Author)	97
39. The end of Freds Flight Alice Springs Airport 1972 (Author)	103
40. Dust storm at Alice Springs Airport (PD Parker)	104

41. Author in the front cockpit of a Tiger Moth Bond Springs NT 1970 (Author)	107
42. Local newspaper reports the day after the crash of Charlie Mike India (Author)	116
43. Beech 65-80 Queenair Aircraft (PD Parker)	117
44. Author taking a bearing on a bush airstrip (Author)	120
45. Simplified Intelligence Plot and witness locations (Author)	129
46. The Department of Aviation RCC at Townsville Airport (Author)	133
47. The radio consolette at Whyalla FSU (Author)	134
48. Author at Whyalla 1973 (Author)	138
49. The radio console at Kalgoorlie FSU 1974 (Author)	148
50. SAR message 1975 (Author)	149
51. Table of SAR Phases (Author)	151
52. Aircraft Movement Message 1942 (Civil Aviation Historical Society)	152
53. DCA FSU at Kalgoorlie 1974 (Author)	155
54. Kalgoorlie Airport 1975 (Author)	161
55. The late Captain Arthur Turner 1981 (Joan MacLeod)	163
56. Kalgoorlie Boulder Aero Club 1974 (Author)	164
57. Memorial in Karrakatta Cemetery Perth WA (Author)	164
58. Kalgoorlie Airport 1975 (Author)	173
60. Flight of VH-FKC Port Hedland to Derby 1971 (Author)	178
61. Fokker Fellowship FK28 jet taking off at Kalgoorlie 1971 (Author)	180
62. Aerodrome diagrams Kalgoorlie before and after 1992 (Author)	182

63. Layer of cumulus clouds en-route (Author)	185
64. Deteriorating weather conditions (Author)	186
65. Author at controls of medevac PN68 Townsville 1986 (Author)	196
66. Post crash pick-up (Author)	206
67. Air Queensland DC3 1985 (Author)	213
68. Townsville Airport pilot briefing office 1985 (Author)	216
69. Author preparing to launch an electronic SAR beacon (Author)	220
70. Aerodrome control tower and communications at Port Moresby 1949	228
71. Mr C.B Hulse operating the Bulolo Aeradio Station 1950's (C.B. Hulse)	236
72. Close up view of the AR7 HF receiver (Author)	239
73. Table of QRM code samples (Author)	243
74. Mr C.B. Hulse operating the Aeradio console at Rabaul 1952 (C.B. Hulse)	247
75. Aerodrome Works DCA MMU Port Hedland 1972	254
76. Pilot Briefing Office Perth Airport 1992 (Author)	264
77. Siemens three-row teleprinter/telex machine (Author)	269
78. The DCA OPS/MET building at Essendon Airport (Author)	270
79. Trans Australia Airlines (TAA) Lockheed Electra 1970 (Author)	270
80. Last message from Alice Springs Flight Service Unit 1991 (Author)	271

Cry from the Kyeema

Captain A.C. Webb watched the compass needle, which was pointing steadily at 110 degrees magnetic. The silver Douglas DC2 registered VH-UYC and named 'Kyeema' carrying fourteen passengers flew its course eastward from Parafield near Adelaide to its destination at Essendon near Melbourne. In fifteen minutes everyone would be dead. Outside the aircraft, two huge radial engines roared in unison.

Webb began to stare through the windscreen at the misty murk. The cloud was continuous at 11,000 feet and he began to ease the aircraft down to an altitude of 7000 feet. He picked up the black radio microphone and called the Essendon Airport Aeradio operator for the weather. The monotone voice of the operator crackled in his headphones. "Overcast cloud at 1,500 feet in the Melbourne area, extending to 4000 feet. Broken layers at 800 feet. The wind is light southerly." There was a brief pause. Webb continued the descent.

> "There's a few breaks now towards the south, down over the bay."
> Webb scanned the area in front of him. The cloud was really thick.
> "Weather received OK," he replied.
> "We're about to enter the overcast at 4000 feet."

The DC2 crew were not happy with the conditions and decided to ask for a radio bearing from the radio station. This could be achieved if they kept the transmitter on for a short duration. But their request was jammed by another aircraft.

They were not to know that in less than half an hour, they would unwittingly be responsible for Government recommendations that would see the removal of Civil Aviation Board as part of the Australian Defence Department and the creation of the Department of Civil Aviation. They would also never know that the creation of this unique Department would oversight the birth of navigation systems, Aeradio stations and Air Traffic Control in Australia.

Outside the cockpit light fluffy cumulus clouds floated underneath the overcast and the thermals began to buffet the aircraft gently. The fourteen passengers gazed out of the window. Some looked at the time. They must be very close to landing soon. The temperature was quite cool but each of them began to sweat in the humid conditions. No one spoke. Hopefully they would soon be on the ground in the fresh air. Webb picked up the microphone to try the radio again. As he did so the ground radio operator called.

"Uniform Yankee Charlie did you call for a bearing?"
"Yes" replied Webb, "what is your barometer please?"
The operator checked his instrument. "Barometer is 29.88...if you want a bearing, keep your transmitter on."

In the cockpit Webb was looking intently ahead. There seemed to be some breaks in the cloud but they were just dark patches. Was that the ground below? Vaguely through the mist he thought that he could see trees. He looked again. The trees seemed very close. This could not be. He glanced at the altimeter now indicating down to 1500 feet. He was not to know that he was nearly twenty miles past Essendon Airport to the east and close to the steeply rising slopes of the Dandenong Ranges. Hidden by the clouds, the mountain loomed ahead. Webb looked out again and this time could see trees directly in front of him in the heavy mist. He grabbed the throttles, slammed them open to increase power and pulled the nose up. Without warning he and his passengers were thrown forward violently amid a horrendous shrieking of destruction. The aircraft disintegrated as it literally flew into the steep slope of Burkes Lookout, a local landmark. A huge explosion ignited the undergrowth.

Back at Essendon the radio operator became concerned at the non-arrival of the "Kyeema" and questioned other aircraft in the area. A horrible silence enveloped everyone. But on the lower slopes of Mount Dandenong some locals near the crash scene already knew the fate of the passengers and crew in the maelstrom of fire on the slopes above. No one could have survived the holocaust. They were beyond any help at all. The local time was a little before 2pm on 25 October 1938.

The author reads the plaque on the long forgotten memorial cairn built 50 metres above the Kyeema crash site on Mt Dandenong Victoria (Author)

The inquiry that was convened into the Kyeema crash was critical of the Civil Aviation Board for considerable delays in commissioning a system of navigation-aids. It was also critical of numerous administrative failures, which included the lack of decentralisation. In May 1939, Prime Minister Menzies appointed J V Fairbairn as the first Minister for Aviation and later in 1939, the new Department of Civil Aviation (DCA) moved into Almora House in Little Collins Street. Fairbairn himself became the victim of an air disaster two years later. He and a number of other government Ministers died when the RAAF Lockheed Hudson in which they were travelling crashed near Canberra on 13 August 1940. The site is still preserved by a memorial cairn with a memorial plaque as shown above, located in the Fairbairn Pine Plantation just over a kilometre to the north of Pialligo Avenue. The fact that the new Minister had been killed in a military aircraft after having recently separated from RAAF control is somewhat ironic. But things were on the move and with the assistance of Amalgamated Wireless Australia (AWA), the Department of Civil Aviation (DCA) began the task of installing communications and flight checking systems and navigation-aids around the country and overseeing pilot and aerodrome licencing.

The Kyeema was finished. The ghosts from that terrible moment, still walk on the slopes of Mt Dandenong. A marker located behind the modern local TV transmission masts stands alone in the forest, testimony to the disaster. But the impact of the accident, the lives lost and the abrupt end of a beautiful aircraft had sent out a cry that changed the face of Australian aviation forever.

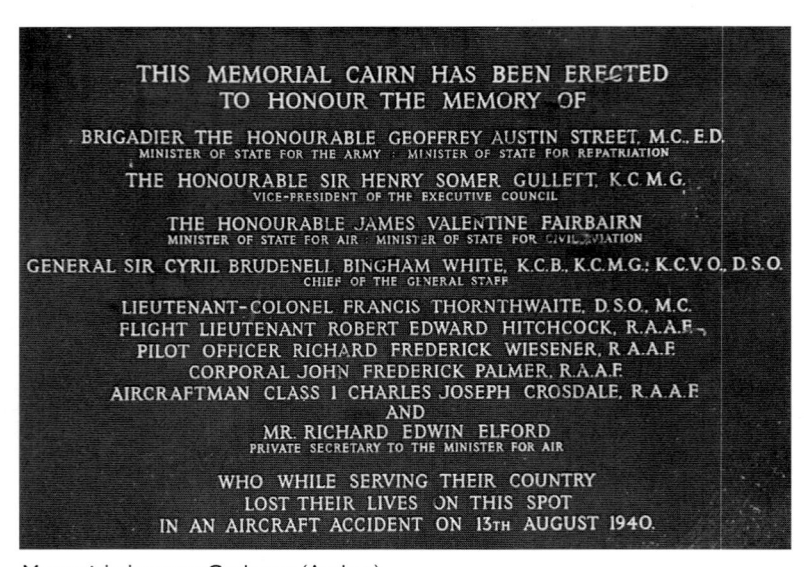

> THIS MEMORIAL CAIRN HAS BEEN ERECTED
> TO HONOUR THE MEMORY OF
>
> BRIGADIER THE HONOURABLE GEOFFREY AUSTIN STREET, M.C., E.D.
> MINISTER OF STATE FOR THE ARMY : MINISTER OF STATE FOR REPATRIATION
>
> THE HONOURABLE SIR HENRY SOMER GULLETT, K.C.M.G.
> VICE-PRESIDENT OF THE EXECUTIVE COUNCIL
>
> THE HONOURABLE JAMES VALENTINE FAIRBAIRN
> MINISTER OF STATE FOR AIR MINISTER OF STATE FOR CIVIL AVIATION
>
> GENERAL SIR CYRIL BRUDENELL BINGHAM WHITE, K.C.B., K.C.M.G., K.C.V.O., D.S.O.
> CHIEF OF THE GENERAL STAFF
>
> LIEUTENANT-COLONEL FRANCIS THORNTHWAITE, D.S.O., M.C.
> FLIGHT LIEUTENANT ROBERT EDWARD HITCHCOCK, R.A.A.F.
> PILOT OFFICER RICHARD FREDERICK WIESENER, R.A.A.F.
> CORPORAL JOHN FREDERICK PALMER, R.A.A.F.
> AIRCRAFTMAN CLASS 1 CHARLES JOSEPH CROSDALE, R.A.A.F.
> AND
> MR. RICHARD EDWIN ELFORD
> PRIVATE SECRETARY TO THE MINISTER FOR AIR
>
> WHO WHILE SERVING THEIR COUNTRY
> LOST THEIR LIVES ON THIS SPOT
> IN AN AIRCRAFT ACCIDENT ON 13TH AUGUST 1940.

Memorial plaque at Canberra (Author)

Doug Whitfield

The Fledgling

I read about the Kyeema crash when I was a teenager. Like many Australians, I was fascinated by the story, especially the accounts of the ground radio operators and how they carried out their daily tasks. I originally developed an early interest in flying because my father had named me after his best friend Doug Burnell, who as a member of an Air Force Wellington bomber crew, had failed to return from a ocean mine-laying mission in World War II. My Father left the Air Force in 1948 and took up employment in the civil aviation industry and as a boy growing up in this environment, I often saw various types of noisy, radial piston engine transport aircraft and early types of jet aircraft flying overhead near my home. But during my last days at school my headmaster had, for reasons of his own, scorned my ambitions for a career in aviation. "Fly?" he snarled, "what do you think you are going to do, grow wings?" He laughed mercilessly and informed me in no uncertain terms that my academic record was not good enough to be considered as a pilot. In doing so, he unwittingly lit a flame of burning passion within me to prove him wrong. Opportunities for an untrained school leaver were few and I was obliged to obtain various jobs to earn enough money to start saving for flying lessons. I purchased an old Volkswagen sedan, so that I could visit the airports in the area. The cry of the Kyeema was beginning to haunt me as I entered the threshold of a new career.

Archerfield aerodrome is near the Queensland State capital Brisbane and became the focus of my full attention. On many occasions I watched enviously, as student pilots performed their "circuits and bumps" on the airfield and I fervently wished that I too, could also fly. In the latter part of 1968, I plucked up the courage to walk into the Royal Queensland Aero Club for the first time and inquire about learning to fly. A fellow called Peter Driver introduced himself as an instructor and enthusiastically discussed how I could begin lessons. Another friendly flying instructor wasted no time in preparing an aircraft for my first familiarisation flight. I walked out of the Aero Club building with Peter and stepped onto the aircraft parking apron for the first time, suddenly feeling a forlorn hope that my dream to be a pilot was about to become true. Peter jumped up onto the left-hand wing of the aircraft. "Hop up here and stay on the black bit –OK?" He grinned and looked up and around at the sky as if making his own assessment of the flying conditions. The "black bit" was a panel painted onto the wing to indicate where to place your feet while climbing in, so that the more fragile sections of the wing were not damaged in any way by heavy weights. I climbed in and gazed around at the cockpit surroundings. There was a distinctive smell of aircraft metal, maybe aluminium. It all looked very complicated. I stifled a worry that maybe I would not cope with such complexities and watched as Peter climbed into the cockpit beside me and fastened the blue webbing harness around himself. "Hows yours?" he inquired leaning over to check that my harness was secure. Words were beyond me and I sat in awe, just smiling happily. I watched keenly as he went through the pre-starting checks. "OK Doug, you can do some of these." He waved his hand around the cockpit indicating that I should do the checks. "Brakes on, Fuel selector on, Master switch on." He paused, "how ya doin', got all that so far?" I hadn't the faintest idea what it all meant and shook my head stupidly. He laughed loudly. "Don't worry, you will wonder what all the fuss was about one day." He turned the fuel pump on and I could hear it ticking loudly in the engine. He pushed the red mixture control in and leaned out of the cockpit to look behind the aircraft. Without warning to me, he shouted "Clear prop!" at the top of his voice. This was to warn others behind the aircraft to stand clear if they were in the way or risk being blown over by the blast of air from the propeller. He pressed the starter button and the engine started nicely and ran smoothly. I could smell the aviation gasoline. "OK" he yelled above the engine noise. "Now we test the brakes," the aircraft rolled forward a metre or so and he tested the brakes. The aircraft bobbed forward slightly and seeming satisfied, Peter started off again. He picked up the radio microphone and said something to the control tower. As he spoke

I turned and studied his face. He appeared to be in his late twenties with dark hair and well defined facial features. His white shirt was impeccably clean and pressed. The gold bars and the gold wings on the left breast of his shirt looked very smart and I gazed momentarily at them with extreme envy. The radio crackled into life as the controller cleared us to taxi and confirmed the weather conditions. The day was very warm and we taxied out from the Aero Club in the little Victa 115 training aircraft and entered a grassy runway. Light winds blew pleasantly across the airfield and presented no problem for takeoffs and landings in these tiny aircraft. Peter quickly outlined the pre-take off checks that are mandatory before departing and smiled at my baffled reactions to these very new complexities before me. He called the control tower again and reported ready. The VHF receiver crackled into life and we were cleared for takeoff. He applied the power gently but positively and we accelerated across the airfield. At fifty-five knots we became airborne and Peter held the aircraft in what is known as 'Ground Effect' until we reached a speed of 70 knots. At this speed we were able to climb safely without fear of stalling the wings in the mild turbulence and at once began an ascent at a rate of five hundred feet per minute. The Queensland panorama unfolded in front of us as we flew over the heat soaked townships shimmering below and the little Victa bounced happily in the occasional air turbulence.

Victa Airtourer V115 two seater trainer (Author)

The sky formed a deep blue canopy above us and we tracked out to the designated flying training area for my air familiarisation. To the east, the Pacific Ocean was clearly visible, sparkling in the summery haze and I began to relax and enjoy myself. Ahead, I could see the pleasure boat canals at Surfers Paradise and on the coast I could see the endless expanse of white surf beaches. To the west, the peaks of the Cunningham Ranges rose to guard the western edge of this lush tropical plain. Peter was now talking constantly to me, explaining the aircraft operation but I was so overcome with the beauty of the flight that I scarcely heard a word. He briefly demonstrated the correct methods of turning the aircraft using a combination of the rudder pedals and the control column. He quickly progressed to climbing and descending exercises and with a mischievous wink, eased the aircraft nose up until the aircraft was pointing at least 45 degrees up into the sky. "Wanna try a stall?" My armpits began to sweat and I nodded dubiously. There was no way I was going to admit that I was scared. The engine was throttled back to lose speed so that the wings began to lose their lift and stall. At the point of stall, the aircraft buffeted and the nose pitched gently forward out of control, into a dive. The horizon ahead disappeared but suddenly bobbed into view in the windscreen again before shooting up above us as the aircraft dived. When the airspeed was sufficient again, the air flowed over the wings at the required speed to regain the lift and we resumed proper flight. I had been holding my breath for a moment and gasped loudly as we pitched forward. I could hear Peter yelling over the noise of the engine and the slipstream. "You OK mate?" It was like a showground ride of some sort. I pulled my face into a silly grin that seemed to reassure him that I was OK. I was feeling very hot and began to sweat profusely. I concentrated hard on everything and there seemed a lot to learn as I sat staring around, trying to absorb every detail. All too soon, we had to return and I sat elated as we dived gently towards the shimmering suburban rooftops and rejoined the traffic circuit at Archerfield. Peter expertly guided the little trainer down the final approach path and we zoomed over the fence past the white gable markers that marked out the airstrip and gently bounced across the green grass to a full stop. He opened the canopy and the breeze cooled my face as we taxied along. Many years later I would remember such a breeze in vastly different circumstances.

We returned to the dispersal point on the flight line, parked next to the other aircraft and shut down the engine. We sat silently for a moment listening to the tinkling sound of the engine as it cooled after its exertions. Peter turned, grinned and climbed out of the aircraft. "Whaddya reckon? Did you like it?" My stomach was still in my throat, but I was certainly happy. I could hardly believe that I had already taken the first step in fulfilling my dream and my walk had the spring of a new born lamb as I returned to my car. Little did I realise that the crash of the Kyeema fifty years before would soon have an invisible but direct impact on my life. Paydays seemed to take forever as I waited eagerly each week for the next flying lesson. My application for a Student Pilot Licence was accepted and the Department of Civil Aviation sent me a copy of the Visual Flight Guide (VFG) and a map of the airspace arrangements around the Brisbane area.

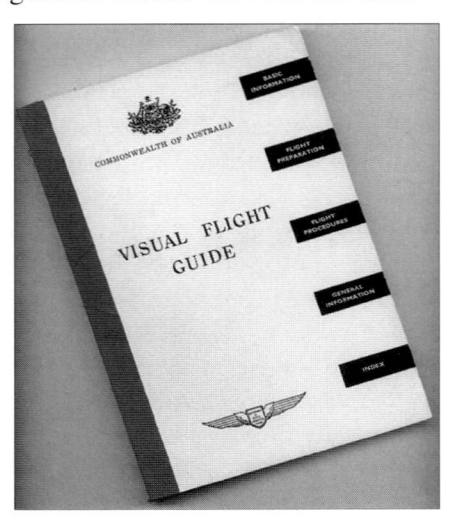

The DCA Visual Flight Guide
From the 1960s era (Author)

These publications were quite bewildering and I spent hours studying them trying to understand all the details. Descriptions of Flight Service communications, controlled airspace's, altimetry, circuit procedures, Air Traffic Control airways clearances, visual signals, morse-code and meteorological forecasting services were just a few of the subjects, that taunted my feeble brain!

By now I had managed to obtain a job at the Amberley Air Force Base as a civilian stores clerk, and this was well paid by my previous standards. During the day I watched F86 Sabre jet fighters and Canberra jet-bombers operating and I yearned to be much more involved with aircraft operations. I decided to apply for aircrew training but failed the entrance test because I did not know at the time that I was suffering from unacceptable short eyesight in my left eye. This was very disappointing, but I was spurred on by the vision of my ex-headmaster telling me that I was not intelligent enough to be a pilot! Unfortunately, I was unexpectedly transferred by the Public Service to the Brisbane Taxation Office in Ann Street because they were short staffed. My position as a clerical assistant apparently rendered me liable for service in any department and this was a terrible blow to me since there were no aircraft in the taxation office!

None-the-less, I worked even harder to improve my income so that I could persevere with my flying lessons. I was eager to learn everything at once and bombarded my instructor with lots of very basic questions. I was puzzled at how aircraft communicated while they were flying well away from the city centres. After all, there were no control towers in the country, so who were the pilots out there talking to? Peter patiently explained how the communications organisation functioned and was operated by the Department of Civil Aviation (DCA). Within DCA was an organisation known as Flight Service, which consisted of a network of radio stations right across outback Australia. Little did I realise, I would soon be intimately involved with this service. As the days passed I noticed that a yellow car was always parked outside the taxation office in Ann Street in the afternoon, when I finished work. The taxation building housed a number of Commonwealth

Doug Whitfield

Government Departments and this vehicle had wings painted on the front doors with the words 'Department of Civil Aviation' painted on a shield in the centre. The wings caught my immediate attention. Surely, this department would be a worthy employer for a fanatical aviation enthusiast like myself! I paid a visit to the office of the Public Service Inspector where the staff produced a set of application forms and a booklet entitled 'A Career in Flight Service.' The description of the course of training closely resembled the details that I had read in the Visual Flight Guide and my enthusiasm was now mounting. I wasted no time in sending off the application forms. The selection procedures were quite demanding, but some weeks later I was pleasantly surprised by a letter formally notifying me that I had been successful and would commence training at the Central Training College in Melbourne in a few weeks time.

DCA Ford Falcon 1970 (DCA)

Early DCA Days

I arrived at Essendon Airport in Melbourne after a long flight from Brisbane in a four-engine Lockheed Electra turboprop airliner. The weather was characteristically cool for Victoria even though it was summer and I stood in my Queensland long socks, short-sleeved shirt and shorts, shivering while I waited for a taxi. Essendon Airport bustled vigorously with noisy aircraft, trucks and trolleys. Even though I was very cold, I stared happily at the sights of the numerous uniforms being worn by local airport aircrews, security guards, kiosk operators and transport staff. A new city is always overpowering to the newcomer and the bustling sprawl of Melbourne left me quite disorientated for the first few weeks. But I had a lot of fun exploring my new surroundings and becoming familiar with the Melbourne trains and trams. In due course I had made a lot of new friends and became totally absorbed in my training. The DCA Central Training College was located in Henty House situated at number 499 Little Collins Street, in the heart of the city. Now, thirty years later, this building has been converted into apartments. In the heady days of DCA, the classrooms were always busy as numerous trainees and instructors moved to and from the various training

DCA Central Training College
Henty House Melbourne 1970 (Author)

sections. The walls of the corridors were whitewashed and the brown linoleum had its own polished odour that permeated the area. On one side of the main corridor there was a classroom equipped with a radio simulator. In the next room was a morse-code training unit.

The classrooms were very cold in winter and students lucky enough to be sitting next to a wall radiator sat in luxury, while the rest of us froze. However, in summer, the windows were often opened and the hum of the Melbourne traffic and bells on the trams could be heard throughout the day. I rented a room in South Yarra and travelled by train daily, to and from the city. My classmates and I used to stand on the railway platforms practicing our morse-code by reading through all the advertising signs we could see on the surrounding buildings! Our fellow passengers must have thought we were very weird as we stood shouting out "COKE!.. dah-dit-dah-dit, dah-dah-dah, dah-dit-dah, dit...." at the tops of our voices.

Back in the classroom we learned that Flight Service Units operating around Australia had pre-determined areas of responsibility and were required to report any aircraft with suspected or actual emergencies to the responsible aviation Rescue Co-ordination Centres (RCC), located in each of the capital cities. The administrative breakdown of the regional areas of responsibility, consisted simply of Papua New Guinea (PNG), Queensland (QLD), New South Wales (NSW), Victoria/Tasmania (VICTAS), South Australia and Northern Territory (SA/NT) and Western Australia (WA) which included the Cocos Islands. The RCCs were usually located at the main airports within these areas. They usually consisted of large rooms equipped with ceiling to floor wall maps, an assortment of drawing tables, rows of telephones and numerous officers responsible for making rapid assessments of the information received from the Airways Operations Units on all operational and emergency matters. The Airways Operations Units located away from the cities usually consisted of Air Traffic Control and Flight Service Units each with its own administrative and technical maintenance staff. These units were governed by various regulatory procedures that were contained in the Air Navigation Regulations, Air Navigation Orders, Air Information Publication and green folders called Airways Operating Instructions known generally as 'AOIs.'

Very often, reported situations resolved themselves fairly quickly and the RCC personnel would then resume duties in the capital city briefing offices and operational control centres, taking flight plans from pilots and giving them updated operational information about weather, navigation aids and aerodrome serviceability. The Operational Control Centres (OCCs), were unique to Australia. The staff usually had extensive aviation experience as Air Traffic Control radar or procedural controllers and also as pilots or navigators.

Most officers knew about the KYEEMA. The impact of the crash had a tremendous effect on the Australian community when the news headlines screamed the details of the disaster, especially since one of the victims was a well known Parliamentarian returning to Canberra from his South Australian electorate.

The major inquiry that was initiated on 28 October 1938 only three days after the accident, continued until December of the same year when several recommendations were made public. Briefly stated, these were that the method of Civil Aviation administration was to be reviewed, that the civil aviation rules and orders become the subject of legislation, that a system of navigation aids be implemented without any delay and that a system of radio and manual 'flight checking' be developed. Application of these initiatives reduced the risks and minimised the many human errors that occurred every day. In record time, the first training course was commenced for Aeradio officers in 1940. These pioneers went on to develop an excellent system over the next forty years and their names were well known by all future trainees

First Aeradio Officers Course 1940
Back Row: G Anderson, E Trebilcock, E Hegarty, L Hall (staff) F Davidson
Middle Row: P Eaton, H Stiff, A Cheney, G Bruce, J Radcliffe
Front Row: M Long, W Matters, E Quilty, O Beckett
Absent: C Carroll, Chief Instructor - PMG Radio Inspector
(Civil Aviation Historical Society)

One of the early Aeradio Officers was Harry Gentle. Harry was one of my early instructors in Melbourne. He once recalled an incident that reflected just how Aeradio stations, were able to provide aircrew with simple, but immediate assistance in the quest for safety. Harry was riding his bicycle to work for the start of a night shift at Canberra. As he pedalled along in the cold night air, he heard an aircraft fly overhead and he spotted it flying in south-easterly direction. Harry was used to the common routes used by aircraft in this area and the direction of this particular aircraft was inconsistent with normally expected operations. When he arrived at work, he mentioned the sighting to the off going officer. The officer stood staring at Harry. "Are you sure?" he inquired. Harry was quite sure he had not made a mistake. The two wasted no time in contacting Sydney Flight Control officers to confirm that nothing was amiss. Swift scrutiny of the flights in the area resolved the matter. An International flight was tracking to Mt Kosciusko at an altitude of 7000 feet and had inadvertently tuned into the Canberra radio beacon instead of the beacon located at Katoomba. The pilot was advised to re-tune to the correct navigation beacon frequency so that they could alter course and head for Katoomba and thus resume his flight-planned track to his destination. If only the Kyeema could have been in clearer weather and enjoyed the benefit of such luck. Harry was typical of the conscientious breed of Aeradio officers who were the fore-runners to Flight Service. He learned his craft in the Australian Army after he left the service to attend the Marconi School of Wireless in 1947. He told his stories in his usual quiet and dignified, almost humble style that seemed to lure me further to the industry that he represented.

Typical permanent Aeradio station room. These stations were located at Hobart, Western Junction in Launceston, Melbourne, Holbrook Sydney Kempsey, Brisbane, Cooktown, Karumba, Groote Eylandt, Port Moresby And Salamaua in PNG. Note the Bellini-Tosi DF receiver and morse Key on the operating table. (Civil Aviation Historical Society)

He and the students that had trained in the early days at the Marconi school, were government sponsored and were required to obtain a First Class Commercial Operators Certificate of Proficiency to graduate. Successful completion of this course satisfied the pre-requisites for employment in the Department of Civil Aviation as Aeradio officers. Once employed, many officers took advantage of the opportunities to serve in many varied and often remote localities.

Harry, typically, was to serve in Canberra, Darwin, Tennant Creek, Mount Gambier, Townsville and Cloncurry. What adventures could be experienced in these remote locations in the quest for providing a safety umbrella for people travelling around the country in aircraft! Harry completed his service as an instructor at Henty House and then as a Regional Supervisor in Melbourne before he retired. I always remembered his neat style of dressing and he used to tell us many stories of his past about aviation incidents. I remember one story he told us as he sat with a smile on his face studying the floor intently as he adjusted his thick framed black spectacles.

"Not long ago, Aeradio stations began to assist with their direction finding equipment in an attempt to assist the pilot find his position. On this particular occasion, the pilot of a United States military transport aircraft reported that he was crossing our northern coastline and was happy because at least he was over land and was sure that it was Australia! Australia was surely big enough to find! Fortunately, when he reached the coastline, he made a right turn instead of a left turn and simply followed it along, until quite by luck he stumbled on Darwin where he landed although somewhat overdue on his original flight planned estimates!. No-one seemed to worry much in those days, but of course in this new age people would become quite excited now."

The excitement emanated from the Kyeema because no-one wanted anything like that to happen again. Elaborate systems were implemented in order to provide the best possible protection for everyone flying. The system of 'flight checking' was developed into two separate services and these became known as Air Traffic Control and Flight Service. Flight Service was originally known as Aeradio and provided search and rescue alerting services and traffic information to aircraft in flight, in areas that did not warrant Air Traffic Control procedures. Air Traffic Control provided similar services but also positively separated aircraft by giving specific directions to prevent collision with each other or with terrain.

The Commonwealth Government contracted Amalgamated Wireless Australia (AWA), to build twelve Aeradio stations. These were to be located at Port Moresby, Salamaua, (also in PNG),

Cooktown and Karumba in North Queensland, Groote Eylandt in the Gulf of Carpentaria, Brisbane, Sydney, Kempsey and Holbrook in New South Wales, Melbourne, Hobart and Western Junction near Launceston in Tasmania. These stations were to provide radio coverage to areas where aircraft were beginning to fly regular air routes between air traffic controlled centres.

A large number of the officers who first manned these stations were mostly ex-marine radio operators who held qualifications authorised by the Post Master Generals Department. In 1939, the Department of Civil Aviation began to recruit staff for 34 stations around Australia and PNG. These officers were proficient in the use of Morse Code, which had begun to be replaced by machine telegraphy in 1949.

I used to trudge off on cold wintry nights to attend radio classes run by some local amateur radio volunteers who were teaching morse-code to aspiring radio operators. We sat for hours listening to our instructor sending groups of letters and numbers on a morse-key attached to an audio oscillator at the front of the classroom. We soon became quite proficient, even a little 'cocky' at our newly found skill, but I was to find out later that you can always find someone better than you. I always admired the old telegraphists. They were dedicated to the task of sending and receiving accurate messages over seemingly impossible distances.

I remember meeting my first telegraphist, not long after my arrival at the Darwin Flight Service Centre. He was working at a radio console, which was known as the Biak circuit, because its primary role was to send weather data in morse code to the island of Biak. Biak is an Indonesian island, which was responsible for air communications in a large area adjoining the Darwin area of responsibility to the north. Meteorological information was sent from the Darwin weather office in the form of pages of sequenced numbers, which represented certain weather codes.

Of course there were no telephones to the islands in those days and the weather office had to rely on a relay service provided by DCA Flight Service and Communications Officers using morse radio equipment. The room was quite small and contained four separate radio consoles. The air was air-conditioned and had a musty smell that is a characteristic of the tropics. Monitor lights winked and blinked on the console and a new smell became evident. The smell of valves and hot radio circuits randomly assaulted my nostrils and somehow I already felt quite at home. I waited politely for the telegraphist to finish sending the message, so that I could introduce myself as the 'new hand' on the station. But he seemed oblivious to me standing there, until I coughed to attract his attention ! He looked up sharply and his face cracked into a broad grin of welcome as he stood up to greet me. As he did so, I was astounded to see that he continued sending the message, never faltering in rhythm, simply changing hands on the key in order to shake my hand with his right hand!

"Gidday young feller, just starting today are you?"

I was amazed. How could he talk to me, change hands and still send morse messages? "I've been doing this for so many years now that its just another language," he chuckled as he jammed a pipe in one corner of his mouth. He adjusted his spectacles while he spoke and his moustache twitched and bristled as he went about his work in a dignified and professional manner. His abilities greatly impressed me and I hoped that I may one day be as good as him. Great care was needed while relaying messages using Morse Code. The radio networks were often operated under difficult conditions caused by interference from electromagnetic disturbances in the ionospheric belt, which surrounds the globe.

An incident occurred on one occasion, involving an international flight that was en-route between Sydney and Djakarta. The flight was forced to divert to Darwin because a message was received that Djakarta airport was closed. Passengers were off-loaded at Darwin and provided with hotel accommodation at the expense of the airline until Djakarta notified that it's aerodrome was open again.

Later, the same flight was en-route to Singapore. The crew contacted Djakarta by radio and asked the Indonesians how long their airport was going to be closed. Imagine their surprise when they were informed that the airport had been open for quite some time! The airport had only been closed for a few hours and not for days as the message received in Darwin had indicated.

Investigation showed that the message, sent in morse had accidentally been sent as the eighth of May and not the seventh. An eight sent instead of a seven had caused a difference of twenty four hours. Commercially, this proved to be a costly error! A checking procedure was introduced, so that further Notices to Airmen, known as NOTAMS included a figure, which provided a check of the numbers received. The procedure was simple. Every six numbers sent in sequence, were added together and the sum included in the message. Thus, if a date time group of 050700 was sent, it would be transmitted as 05070012. If an operator received the group as 05070013 an error was obvious because 5 and 7 do not equal 13. A request for confirmation would be sent back to the originating station and they would send out a corrected copy.

Needless to say, the communications officers became very good at arithmetic! A photograph of my graduation from the Flight Service Course arrived in due time and I looked forward to my new career in earnest.

FLIGHT SERVICE COURSE No 11 JULY 1970
Rear left to right trainees: Ron Chamberlain, Bob Burrows, Phil Williams, Viv Saal, Kev Carter, Ron Dunn, Mike Hayden, Dave Jones, Les Anderson, John Ruciak, Doug Whitfield, Jim Buckley
Front left to right Bob Chapman (Ops Instr) Ray Acaster (Met Instr) Harry Gentle (OIC retiring) Sid Hulse (OIC elect) Don Brown (Nav Instr) Col Peacock (trainee)

First Bush Posting

I arrived at Katherine in the Northern Territory late in the evening aboard a Fokker Friendship FK27 Turboprop aircraft operated by Ansett Airlines. I noted that it's registration was VH-FNI and this was the very aircraft that was the subject of the first known airliner hijacking in Australia at Alice Springs two years later in 1972. The Katherine township is located approximately 300 kilometres southeast of Darwin on the Stuart Highway. This highway is the road communications link that connects Darwin with Alice Springs through the dead heart of the Australian continent. Katherine was the centre of a beef cattle region and the meat works near the town processed beef for export. Sorghum and peanuts were grown on local farms and a large CSIRO research station nearby investigated agricultural processes. The Katherine River flowed through the town and a few kilometres upstream the Katherine Gorge attracted numerous visitors from all over the world and became a very popular local tourist attraction.

The aerodrome at Katherine was very old and no longer suitable for the new modern range of turboprop aircraft that was replacing piston-powered aircraft used by the airlines. The runway was quite weak with a low pavement classification number but the Connellan Airways De-Havilland four-engine Heron DH114 and Northern Territory Aerial Medical Service, twin engine, De Havilland Dove DH104 aircraft could still use it quite safely. Larger aircraft were catered for, a few kilometres to the south of the township at a new airstrip constructed for Australian military exercises. A new VHF Omni Range (VOR) and Distance Measuring Equipment (DME) radio-navigation aids served the bitumen runway. This new facility had been officially named Tindal in memory of Wing Commander Archibald Tindal. Tindal was the Area Armaments Officer at the Darwin RAAF Base on the 19 February 1942 when a large Japanese bomber force led by the famous Pearl Harbour attack leader Commander Mitsuo Fuchida bombed Darwin. A radio warning was received by the naval communications station at HMAS Coonawarra near the Darwin RAAF Base as the Japanese approached the coast. Fuchida led his force towards the initial attack turning point over Noonamah to the south of Darwin before turning north on the run towards the still unsuspecting city. Inexplicably, no alarms had been sounded and as the bombs rained down, the RAAF station staff ran to operate the ground defences against the air attack. Wing Commander Tindal was one of several airmen that ran to operate machine guns from the air station trenches as fighters tried to scramble, but his position was spotted and strafed by the enemy fighters. He died bravely, trying to defend his country against a superior fighting force and now lies at rest with many other service personnel at the Adelaide River War Cemetery not far from Darwin.

There was no RAAF base at Tindal in the 1970s and the airstrip was used specifically for jet operations during military exercises and also for passenger services run by Ansett Airlines and the government owned Trans Australia Airlines (TAA). Later it was to become a fully operational fighter base for the RAAF. When I first arrived there in 1970, there were no facilities at the airstrip except for a temporary corrugated-iron shed that was used for an air terminal. The shed was constructed so that arriving passengers could be accommodated and others could be processed for departure in some artificial shade. The temperature and humidity was always very uncomfortable and sticky in the wet monsoonal season. It was late in the evening and the deepening vestiges of the red sunset had long merged with the overhead panorama of stars.

I waited patiently while baggage was offloaded from the FK27 onto battered and rusting trolleys. Soon all the baggage had been seized by its owners and they had vanished very rapidly into the night in an assortment of vehicles that had been waiting to transport them into Katherine Township. However, even though I had been watching carefully for my bag, it had obviously not been unloaded from the aircraft. When the pilots closed the aircraft and made ready to depart, I watched in dismay as the left hand engine whined into life and disrupted the peace of the evening again clearly intent on keeping my bag on board. The local airline agent had already disappeared, so there was nobody to ask for help. Annoyance overtook me and I ran out onto the tarmac to a position where the pilot could see me clearly in the floodlight and literally flagged the airliner down! I could see him looking down at me with a puzzled expression on his face. Behind him I could see the co-pilot half standing out of his seat on the other side of the cockpit so that he could get a better look at me and the two pilots seemed to be in earnest discussion. After a few minutes, the engine nearest to where I was standing, was shut down and as soon as the large four bladed propeller stopped turning, the cargo hatch opened. The co-pilot came tumbling out and ran over to where I was standing. He asked me what was wrong and I told him that my bag was still on his aircraft. I showed him my ticket with the bag label stapled on the front. He was not convinced and suggested that perhaps someone else had inadvertently taken it. I knew that my bag had not been taken off the aircraft and insisted that we at least have a look on board. With a sigh of despair, he helped me aboard through the cargo hatch and we both began to ransack the aircraft luggage compartment. The passengers gazed on with disinterested stares as I gleefully found my bag at the bottom of the compartment. I cheerfully thanked him and hopped off the aircraft onto the tarmac again. I stood watching from a safe distance, as the engines were re-started. With a wave, they taxied out and took off into the night. As the roar of the engines receded into the distance, I picked up my bag and wandered around to the other side of the terminal shed, only to find that I was now completely alone. I looked for a telephone to call a taxi, but there was no public telephone anywhere to be seen. There were no houses nearby and the area was very still, except for the "chirping" noises being made by a variety of tropical insects. I knew the highway was close by, because I had heard the occasional vehicle going past, so I set off down the dirt track on foot, carrying my bag in the hope of hitching a lift into town. I knew that Katherine was approximately 11 kilometres away and decided that if I could not get a lift, I would simply walk all the way. This was a rather demoralising prospect so late at night. My hopes soared as a battered Volkswagen Kombi vehicle came into view with only one headlight working properly. But to my dismay the driver completely ignored me as he hurtled past, showering me in a cloud of thick red dust. The track went nowhere except to the terminal and I knew that he would have to return this way. I swore under my breath and turned to follow him back to the terminal. But then I paused. If he had not stopped the first time, it would be quite likely that he may not stop the second time, in which case I would have walked all the way back to the terminal for nothing. Dejectedly, I succumbed to my own logic and turned back into the night, heading for Katherine.

The tropics exudes its own unique smells at night and I could particularly smell the gum trees tonight. The darkness of the bush at night alerts your senses dramatically and needless fear prevails in the minds of those who are fearful. I was quite fearful, though I did not know why and jumped with fright every time something moved in the bush. I fervently hoped that the movements were only being caused by harmless creatures like lizards or mice!

I heard the Volkswagen Kombi start up again back at the terminal shed and the noise of its clattering engine drifted easily through the bush to me. I listened as it came closer, heading back down the track from the terminal. With a grinding of gears it reappeared and stopped at my side, showering me in red dust again. The driver, was a swarthy looking, heavily built man of Mediterranean origins. He leaned out the window and informed me that, "Tassie (Tasmania) was back the other way mate!" Such was Territorian humour. He offered me a lift and we drove together in the night with little conversation. One kilometre from Katherine, he turned off the highway down a dusty bush track that led to one of the nearby cattle stations. He told me that it was too late to drive me right into town and I wondered vaguely what I had done to upset him. I was too tired to argue, but was soured by the thought that the circumstances only justified a detour of one kilometre to drop me off. Once more, I trudged off into the night and eventually reached the main street in the town where a somewhat inebriated local staggered into my path on his way home. I stopped him and asked for directions to my lodgings, but he could only totter about in the road, wave his arms in a haphazard fashion and mumble that, "I had to go down to the other end of town." As I set off again, a native woman lurched out of a nearby alleyway and ran after me. Reeking of alcohol, she made noisy demands to the effect that I should father her children. My nerves finally snapped and despite my exhaustion, I picked up my bag and with a Herculean effort, fled down the Katherine main street at a better than Olympic standard. By luck, I found my lodgings and banged miserably on the front door until the landlord signed me in with less than good humour and handed me my room key. I locked my door and sank wearily onto the bed where I fell asleep until late the next morning. My posting to Katherine DCA had a few more surprises in the weeks that were to follow.

I was in no way prepared for the unique lifestyle in this region. The DCA training college had not exactly told me about living in the bush and I found my first exposure to the rough and tumble of the northern end of the Northern Territory very interesting. The local people seemed quite lethargic and very direct with everyone, despite their social differences.

"No bullshit 'ere matey" I was told by a local barman. "Tell it 'ow it is, or bugger off, orright?" Having just worked very hard to graduate from the DCA Central Training College I was not of a mind to "bugger off," and so I decided I had always better make sure "I told it how it was," whatever that meant.

The Flight Service Unit at the Katherine Airport was only three to four kilometres out from the town along the Katherine Gorge tourist track. The Katherine hospital was located directly opposite the airport between the track and the river. At one end of the single runway was the town cemetery and local humorists were quite adamant that aircraft arriving with any kind of emergency, should land from the opposite end. This was so that if the pilot lost control, he would most likely finish up digging a hole in the cemetery end of the airstrip and thus save everyone from burial tasks!

Unfortunately, the humour was based on stark reality. On the 18 January 1939, only two and a half months after the Kyeema disaster, a Guinea Airways Lockheed Electra Model 14 registered as VH-ABI and named Koranga crashed on take-off into the

Flight Service Officer Don Hodder operating the Katherine VHF and HF Radio Air/Ground/Air network at the Katherine Aerodrome October 1970 (Author)

Katherine River not far from the Katherine hospital. There was no Aeradio station at Katherine Aerodrome at that time and the aircraft had made a scheduled stop there while en-route from Darwin to Adelaide. The local Guinea Airways agent met the aircraft and there were only four occupants on board including the two pilots and an engineer. A heavy rain storm began and the pilots decided to delay the flight until the storm had passed by. During this time, they used their HF radio to contact a network Aeradio station, probably at Daly Waters or Darwin to report their situation and intentions. When the storm had passed Captain Clarke taxied out with a senior company pilot, Captain J. Jukes sitting in the right-hand seat and began to take-off towards the Katherine River. Doctor Clyde Fenton, well known for his aviation exploits had noticed the flight was late departing and this had caught his interest, although he thought that the crew had probably waited for the weather to clear. He and several other local people paused in their daily routine to see the aircraft take-off. The angry roar of the twin radial engines increased as the aircraft sped along the runway towards the onlookers. But something seemed wrong. The aircraft did not seem to be accelerating and the nose seemed to pitch up and down. Everyone watched in horror as it became airborne right at the end of the runway in an unusually high nose up position, just clearing the airfield fence. It floated straight ahead towards the trees on the banks of the Katherine River. Impact seemed imminent but the aircraft banked to the right and appeared to be trying to follow the river, which was a little lower than the surrounding terrain. The angle of bank at a speed lower than flying speed increased the drag of the airflow and the aircraft side-slipped to the right and disappeared. At such a low speed and angle of bank, the aircraft very likely lost its lift, stalled and crashed. All the on-lookers heard the crash and rushed at top speed to the site to try and help the victims. There were no survivors. A telegram sent from the Aerodrome Inspector at Darwin to the Civil Aviation Offices in Melbourne advised of the terrible basic details, including the fact that the only passenger was another Aerodrome Inspector. I was amazed that no memorial was evident in the township and worse, only a few of the locals could tell me anything about the accident.

The airport had developed a little more since then and consisted of a small terminal for the local air charter companies, the Flight Service Unit, a groundsman's workshop, a house for the groundsman and a small barracks for the single male staff. The barracks consisted of four small bedrooms each equipped with a rusty iron bed and battered wardrobe. A wooden veranda connected the rooms with a kitchen at one end where meals could be prepared privately. Unfortunately for me, this accommodation was already full and so I was forced to live in lodgings in the town. During one week however, my lodgings were renovated and there was literally nowhere else to live and so I simply slept, dressed only in my underpants, on the briefing office counter at the airport and lived on cups of tea, beer, biscuits and salad sandwiches. At least the briefing office was equipped with a wall-mounted air conditioner, although on one hot night the cleaning lady arrived somewhat unexpectedly and we both got a fright! She seemed to become quite friendly after that!

Katherine Flight Service was an interesting Unit in which to work. The airport was not especially busy, but a wide variety of aircraft types arrived and departed each day. On one occasion, I was standing by the fence near the aircraft-parking apron talking to a friend. We idly watched a Grumman Agcat cropduster aircraft taxi in after landing. Its yellow and blue paintwork had faded from carrying various chemicals in the hopper but the propeller was new and shone in the sunlight. The Pratt and Whitney R985 450 HP radial engine roared magnificently as it swung around in front us and then coughed to a stop as the pilot shut it down. He climbed out of the aircraft and wandered over to where we were standing. He was a lanky individual wearing blue jeans, a torn cowboy shirt and dust covered cowboy boots. His face was grimy and his hair was sweaty but he was friendly enough.

"Gidday," he grinned. "Anyone goin' inter town?"

We each shook hands.

"No mate, my car has busted, guess you'll have to walk."

An old airfield worker nearby overheard our conversation and ambled towards us.

"I can give 'im a lift mates, no worries."

The pilot smiled and accepted the offer. The old man pushed his battered bush hat onto the back of his head and leaned against the fence.

"Yeeup, will be my pleasure I'm sure," he murmured. His whiskered face smiled mischievously showing his broken and yellowed teeth as the pilot walked quickly back to the aircraft to retrieve his overnight bag from behind the pilot's seat. There was no room next to the driver because the passenger seat was occupied by a very large and grumpy bull-terrier. The pilot jumped into the tray at the back and they all disappeared down the airport road towards town.

My friend turned to me with a quizzical look.

"That old guy was up to something devious don't yer reckon?"

I nodded, but guessed it would not be long before we found out, after all, this was only a small town.

Communications with aircraft landing and taking off from

stations and settlements in the flight information area of responsibility meant that the radio frequencies were extremely busy while pilots made their reports. Most pilots chose to proceed on what was known as "full-reporting" radio procedures. This required them to report when they taxied at the departure point, report departing and then make reports at positions they had nominated on their flight plan until they arrived at their destination. On arrival they were required by law to report a safe arrival. Katherine FSU Flight Information (FIA), encompassed the northern end of the territory from east to west. However, this area was cut off by a line running east-west through Pine Creek on the Stuart Highway to the north and another line also drawn east-west through Daly Waters on the Stuart Highway to the south. Daly Waters had ceased to operate as a Flight Service Unit because its area of responsibility had been taken over by units at the Katherine and Tennant Creek aerodromes. Other adjoining FIAs were operated by Wyndham FSU to the west and Mount Isa FSU to the east. If aircraft crossed these boundaries, their last position would be relayed to the next FSU, who would then accept responsibility for that particular aircraft until it either, flew out of their area or landed safely within that area.

Two officers were normally on duty at Katherine airfield from 0500am until 1300pm and these two officers would relieve the morning shift with a sixty minute overlap from midday until 2000pm. They were all qualified to perform the full range of duties, but on each shift at least one operated the radio console, while the other performed briefing office duties. The briefing officer also prepared notices to airmen (NOTAMS) and performed weather reporting duties every thirty minutes for the Bureau of Meteorology. Weather officers from the Bureau were not employed at all FSU locations. The Officer in Charge (OIC) was Don Middlemiss and as OIC, he normally performed administrative duties in an adjoining office from 0800am until 1700pm and also worked certain rostered hours on one day per week, in order to maintain proficiency in communications operations. My car was still unserviceable and off the road for a week being repaired. Unfortunately, two of the other Flight Service Officers suffered

Northern Territory Aerial Services Medical De Havilland Dove registered VH-DHN parked on the main apron at Katherine in October 1970. The Flight Service Unit is directly behind the aircraft. Note the Civil Aviation Ensign flying from the mast. Also visible is the meteorological enclosure on the left of the photograph. (Author)

The disused Katherine Flight Service Unit in 1997 now used as a museum for the local township. Sadly, the airfield is closed and being redeveloped for housing and the advanced deterioration in the apron surface conceals the romantic age of aviation that has now passed into history. The windsock has been erected for the occasional use of a local helicopter operator. (Author)

from vehicle breakdowns and so, rather than impose myself on various wives who began a rostered ferry service to and from the town, I decided that I should look after myself and walk instead. One afternoon, I finished work and set off from the airport to walk into town. The temperature was unusually hot for the tropics and was recording 39 degrees Celsius. By the time I had walked into the main street of the town, I was desperate for a drink and casually entered the bar of the Commercial Hotel. The hotel had seen better days and was about to be demolished and rebuilt. The bar in 1970 was like a prop from a wild-west movie. I bought a beer and leaned on it trying to merge with my surroundings but dressed in my public service airport rig, I stood little chance of looking like some of the grubby station hands that were hanging around. There by the bar was the old airport worker that I had met a few days before. He raised a hand to say hello and I nodded in return. "How did you go with the pilot the other day, I hope he shouted you a beer?"

He grinned and I noticed that he had not changed his clothes from when I last saw him at the airport. "Yeah mate he did." I ordered a beer and settled myself on a bar stool.

"You were up to something you old blighter, what were you doing?" He looked at me disdainfully. "See here young feller. That bloke spends all 'is time sittin' in 'is plane behind a tonne of bug-spray and other stuff." "So?" I asked. The old man sipped his beer and stared at me very seriously. "Well, 'e sits there all the time soakin' up the stuff and after a while he can't get it of himself can 'e?"

I zapped a mosquito that had just taken a liking to my neck.

"See?" laughed the old man.

"No I don't see, what are you on about?"

"Mossies, flies and other bugs all around the bar eh?" He waved his hand around. "Well, when a croppy (cropduster) like 'im comes inter town he pongs like a tin of mossie repellant doesn't 'e? So if you buy him some beers, there's no mossies and bugs in the bar fer the night, right?" I laughed and the old man stood up, pulled his hat over his eyes and ambled outside onto the street. Today was pay-day and the station-hands were in town for a night out. As I was calling the barman to order another beer, a sudden movement nearby, caught my attention. I was just in time to see one of the station hands rushing towards me with a bowie knife raised in his hand! I jumped smartly out of his way and was relieved to see that his target was not myself, but another station hand who was standing on my left. They were from rival cattle stations just out of town. One of the combatants was aboriginal and the other a white Australian. If the fight had been between the two races of people then they would have been easy to identify, but unfortunately, there were members of each on both sides! In only a few moments the pub became a frenzied sea of arms and legs as the opposing sides engaged in a glorious punch-up. I decided to escape while I was unscathed but the fight had moved between the only exit and me. I tried to push through the crowd, but was soon being punched and pushed around. Soon, I was in a lot of pain from the blows and had even thrown a few of my own punches back at the offenders. I was obviously not going to get out unscathed and so in unfettered cowardice, I dropped to my knees onto the floor and crawled out on all fours through the sea of legs to the exit. I arrived on the footpath outside very dusty and bruised. I heard a chuckle nearby and there leaning against a police wagon was two very amused Northern Territory police officers. I tried to explain myself thinking that I was about to be arrested but was relieved to find out that they were not really all that interested in me. I thought it was rather odd that they were making no attempt to break up the fight and asked if they were waiting for more police to turn up. They laughed and told me that the "next nearest copper was in Daly-bloody-Waters mate" a long way down the track and since they were the only police available they were quite happy for the fight to burn-out of its own accord. In any case since they were hopelessly outnumbered there was little point in getting beaten-up for no reason. It was not long before battered contestants from the fight came stumbling through the pub doors rather like a Hollywood cowboy movie. While the disorientated fighters staggered about on the footpath, the two policemen quietly arrested them and locked them in the rear of the police wagon. When six or so fighters had been apprehended, the van drove off to the town lock-up, to drop them off before returning to pick-up another load. The process continued for an hour or so until everyone decided that enough was enough and dispersed into the evening leaving the publican to clear up the mess. I found out later that the police did not bother with formal arrests. It was easier to let the offender's sleep-it off in the watch-house and then go home in the morning. After all this tropical madness, I decided to ask one of my co-workers for a lift to the airport and back until my car was repaired. This would control any urge to go to the pub for a drink again. I figured that maybe I would not be so lucky next time! Regular meals in Katherine were difficult to obtain unless you did your own cooking. The exception was a local diner on the main street. For a reasonable price at Petersons Cafe, single residents could purchase a two-course meal and a mug of tea. The café was very old and somewhat run-down in appearance but typical of most of the eating places available in outback towns at the time. Large blow-flies flew in huge numbers around the counter and tables because there was no fly-wire on the doors and windows. An old air-conditioner rumbled in the corner, but its days of actually properly cooling anything had long gone. Most of the customers were like myself, new in town and single. The heat of the cooking stoves made the air almost unbearable and customers sat at their ricketty tables sweating profusely. The meals were good and nearly always consisted of buffalo steak and salad, but the second course varied and quite often featured apple pie and ice cream which was highly attractive to me, despite the heat. One lunch time on a day off, I was sitting in the cafe when two of my DCA workmates spotted me sitting there. Leaning in the doorway, they announced that they were going down to the 'low level' for a swim with their families. They told me that I was welcome to go along too if I wanted to get my swimming kit and go right now. It was hard to resist. The day was very hot and there wasn't much to do. The only problem was, I couldn't swim! I could doggie-paddle after a fashion but not swim with any confidence. There were two bridges over the Katherine River at opposite ends of the town. One near the main street was referred to locally as the 'high level' and carried the railway line that operated between Mataranka and Darwin. The Stuart Highway from Darwin, or 'the bitumen' as it was known locally, actually crossed the river further downstream, where the gradients of the river banks were not so steep. This point was known as the 'low level,' and was also a popular swimming spot. Unfortunately, crocodiles were known to pass through from time to time and a careful watch had to be kept to ensure safety especially since a nearby abattoir apparently attracted these reptiles. I was somewhat sceptical about the authenticity of the existence of the crocs, but had no proof to the contrary. For a non-swimmer to enter such waters, gives an indication of just how hot the weather could be in this location, but urged on by my workmates, I entered the cool and refreshing water. The group consisted of local pilots, DCA Flight Service Officers, airport technicians and groundsmen, so I felt quite at home with them all. I paddled about in the shallows quite happily, but soon became acutely aware that even the smallest of children belonging to my workmates could swim very well indeed. Their amusement at my inability to swim properly, began to make me feel rather foolish and I started to wish that maybe I had gone flying instead. They insisted that I jump off a rope hanging off a nearby tree over the middle of the river but I declined, feeling a stronger need for survival! My friends sat about in the bush talking and drinking from cans of beer and every so often would jump into the river to cool off, swimming to and from the opposite bank. In between each swim, they sat on the grass, sipping another can of beer and loudly provided me with advice on how to perform the best swimming strokes. Before long, they decided it was time to do their duty and teach me to swim. My embarrassment was mounting as I was towed out into the deep water for lessons. A large audience watched from the embankment and shouted encouragement at my pathetic attempts to swim. Suddenly, a shout from the onlookers warned of an approaching crocodile and my instructors deserted me with amazing speed to regain the safety of land. I was left in the middle of the river, wide eyed, frantic, spluttering, swallowing lots of water and struggling violently. Fearing the worst, I ducked my head into the water and rotated my arms like a demented windmill, hell-bent on survival. I struggled up the embankment, gasping madly for breath, spitting mud and grit. My friends no longer seemed alarmed and had resumed their socialising nearby. I yelled at them to find out where the crocodile had gone. One of my workmates named Leo, came over and banged me heartily on the back. "What croc was that mate?" he grinned. I spat more mud and queried again what on earth was going on.

Leo and the others roared with laughter, "There's no croc mate, that was yer first swimming lesson!" He waved his beer can in the direction of the water.

"Look, you went from over there, to all the way over here." I stared dumbfounded. I'd been tricked into learning to swim. They threw me back in and I had indeed, learned to swim. My posting to the DCA Darwin Flight Service Centre arrived and the following week, I boarded VH-CLT, a Connellan Airways De Havilland Heron DH114 at Katherine and continued my adventure, further north.

Connellan Airways Heron DH114 registered VH-CLT parked on the Katherine Aerodrome Apron November 1970. (Author)

Darwin Duties

Darwin Airport re-opened its Aeradio operations after WWII, and commenced providing updated radio operational information messages and search and rescue alerting services to aircraft flying on both domestic Australian internal flights and on popular International routes destinations in South East Asia and the Indian Ocean.

One of the first officers to start working in Darwin was Lindsay Telford who joined DCA as a communications officer in 1949. He served in radio units at airports around Australia, operating from Essendon, Darwin, Launceston, Broken Hill, Dubbo, Sydney and Melbourne. Before he retired, he served as a Flight Service Instructor in the Central Training College and became a Regional Supervisor in Head Office in Melbourne before handing over to Harry Gentle. He first arrived in Darwin in 1950 and bought a house at Nightcliff on the north coast for an amount of 250 pounds sterling in the old currency, before the arrival of the present day decimal dollars and cents. The house consisted of a converted WWII army hut, that had been vacated by the US allied forces that were formerly stationed in the Darwin area. The building was constructed of concrete, with a black iron roof. Air conditioning could only be achieved by the use

Lindsay Telford on duty in the Darwin Aeradio Station communicating with aircraft using a morse code key. Taken in 1950. (L Telford)

of an electric fan but unfortunately, the power supply was provided by an old WWII generator. This was usually shut down between 6am and 6pm during the day in order to conserve power and prevent overheating of the flimsy equipment during the heat of the day and did nothing for the comfort of the DCA officers and their families.

In fact conditions were quite oppressive for the wives and families and certainly did not help officers trying to sleep after the night shifts. Lindsay recalled that conditions were quite harsh and supplies were difficult to obtain. Butchers in Darwin supplied meat dripping with blood and wrapped in old newspaper. There was never any choice of meat and you simply had to take what was offered or simply go without ! Like many of the operators of the time, Lindsay learned his communications skills in the army as a signaller. He trained in coding and encoding of signals and while assigned to a special wireless interception unit, he learned the KANA code in order to successfully intercept Japanese morse-code traffic. This was a formidable task indeed for non-Japanese speaking Australian operators! The message traffic was usually transmitted to other stations by using Morse code and individual operators developed their own style of sending messages. These styles soon became recognised by other operators, so that individuals were always identified by the particular pattern of their transmission style. The identifiable differences of styles between the radio operators using Morse-code techniques became known as the officer's 'fingerprint' and even though direct voice communication was not employed, operators found no difficulty in identifying each other. One night in Darwin, Lindsay recalls that he was monitoring the progress of a Qantas aircraft in transit between Singapore and Perth. All the position reports were transmitted in morse code by either the on-board radio officer or the navigator. On this particular night, Lindsay suddenly recognised the Morse-fingerprint of an old friend, Ken Wolenski. He quickly transmitted, "is that you Ken?" to verify the identity and also let Ken know that Lindsay was on the radio circuit. Wolenski instantly recognised Lindsay's particular style and responded with a quick "Yes Linds," even though no direct voice transmission had been made. This rather unique method of identification converted an otherwise impersonal method of telegraphy into a pleasant social camaraderie.

The long trans-continental flights in the old and slow airliners of the day could present hours of boredom for the crew, a time interspersed on occasions with some moments of fear when little things went wrong. Nothing could be worse halfway between landmasses and over the ocean when an engine started to malfunction. Pilots therefore relied on the network radio stations for security and safety. A quick call to the radio station would alleviate the feeling of isolation and everyone on board would rest a little easier in the knowledge that someone on the ground was sharing a particular problem and able to send help in a critical situation.

During the London to Australia air race of 1954, one of the competitors was flying an old WWII light twin engine Mosquito bomber. Unfortunately, they became lost between Singapore and the Cocos Islands and radioed a distress call. Aeradio operators located in both Singapore and Port Hedland in Northwestern Australia, picked up the distress call and began to provide continuous radio transmissions so that the pilot could home in on their station signals. By doing this, he could find the bearing to the ground station by turning the loop antennae on the aircraft and find a position line that could be plotted with reasonable accuracy on a navigation chart. He could then quickly repeat the procedure using another ground station and plot another position line on his map. Where these two position lines intersected was very close to his actual position and thus he could relocate himself and fly a heading from that point to a known destination. Of course two position lines would not produce a very accurate position because the aircraft was moving fairly quickly. The bearing from the ground station was also subject to certain position errors and radio signal variations. This was not a good day for this particular pilot. He was able to find the Singapore coastline dead ahead of the aircraft but ran out of fuel just before he could reach it! He ditched his precious aircraft in the sea and finished the race by swimming ashore and reporting the matter to somewhat startled race officials.

Sometimes, communications were interrupted by power failures at the ground station. Lindsay Telford was in Darwin one night when a small cyclone hit the town some twenty years before the terror of Cyclone Tracy destroyed everything in 1974. Lightning hit the Darwin Aeradio station blowing fuses across the room in a miniature but none-the-less impressive fireworks display, which did little for the morale of the officers on duty and plunged the station into an eerie silence and darkness. Lindsay remembers getting a very nasty scare and had to personally proceed along half a mile of bush track in torrential rain and high winds to locate the emergency power generator hut, which served the station. After a short while in somewhat distressing and dangerous conditions, he was able to reconnect the power supply and restore the station to normal operations. Nowadays, thankfully, all radio stations and air traffic control units are equipped with a modern uninterrupted power supply unit, known generally by its mnemonic 'UPS.' Staff shortages were severe during the post war years and officers often worked seven days per week for considerable periods of time to ensure the aeradio station provided a continuous twenty four-hour service. This required great dedication to duty and no thought was ever given to modern day union attitudes.

When I arrived at Darwin during a monsoonal afternoon in 1970 on board a Connair Heron DH114 we landed on runway 36. Little did I know that this would be the runway from which I would fly my first solo in the following year. The November air was already very humid as the heavy rains had already started in earnest. Despite a thunderstorm raging over the sea to the north, the local weather was temporarily sunny with a many fluffy cumulus clouds forming around the town and aerodrome area. After my first solo flight from runway 36 it always seemed to represent hallowed ground in my later years.

I was accommodated initially in the Commonwealth

employee's hostel on the Esplanade in the city but not long afterwards moved into the Ross Smith Commonwealth Hostel in Parap where the accommodation huts had just been taken over from the RAAF. The old DCA hangar was still being used as a vehicle maintenance workshop and across the road was the Darwin Aviation Club (DAC).

Nearby, the Commonwealth Government was constructing two large three story single persons accommodation blocks to replace the RAAF huts. I put my name on the list for one of the new rooms as soon as they became available. The hostel featured an old fibro building that was used as a mess hall and meals were provided army style, from a servery in the dining hall. The meals were always good although many complained about them and I used to wonder what sort of meals they were used too at home if they were grumbling about the food provided in this establishment. For the most part however, camaraderie amongst the mixed tenants was excellent and moral was very high. I shared accommodation with Airport Fire Crews and technical staff from the Postmaster Generals (PMG) Department. We used to keep fit by running along the Mindil Beach road and swimming in the Parap Public Swimming Pool. There were frequent sporting venues and I played Baseball and Cricket with the Waratahs Club for some time.

One of the disused RAAF huts being used as accommodation for Commonwealth Employees at the Ross Smith Commonwealth Hostel in 1970. The new Parap public swimming pool can be seen in the background. Photo Author

When my DCA training was nearly complete, I was very excited when I was awarded my first operational field rating and this meant that I had finally graduated as a fully qualified Flight Service Officer. I was posted to the staff at the Darwin Flight Service Centre. In the centre, the Communications consoles were arranged one behind the other along one side of the radio room. The smell of hot radio components and aluminium casings pervaded the air. Each console was operated by a Flight Service Officer wearing a set of headphones equipped with a boom-style microphone. When air traffic conditions became busy, two operators were required on the console. When two operators worked together, one operated the radios while the other operated the teleprinter taking the aircraft reports direct to the keyboard. Typing speeds therefore had to be well in excess of 40 words per minute without errors! The consoles were also equipped with a larger boom type microphone and loudspeaker so that during low traffic conditions, the operator could walk away from the console to perform duties such as checking teleprinters or completing daily traffic records, but still able to hear any aircraft calling. Every console was equipped with a myriad of lights illuminating the various frequencies being used and the high frequency radio static could be heard even when operators were wearing headphones.

The Darwin Flight Information Region Operations Building which comprised an International Briefing Office, Operational Control Centre, Meteorological Office, Flight Service Centre and Communications Centre. These units have now unfortunately been disbanded and this building was demolished in 2000.
Photograph taken 1997. (Author)

Teleprinters clattered in the background, pouring a continuous stream of weather information and flight planning data onto printed rolls. Paper message tapes ran from the receiver units so that all messages could be re-transmitted on a relay basis if required on the same teleprinter or on the transmit facility on another teleprinter.

In an adjoining room, briefing officers worked on 24-hour rosters discussing operational information and weather details pertinent to the safety of flight with aircrew. On completion of the complex briefings they accepted flight plans from the pilots and prepared them for transmission to all air traffic control and flight service units, responsible for the areas in which the aircraft would be flying. The briefing officer normally accepted a copy of the flight plan from the crew and stamped the date and time of acceptance on the bottom right-hand corner. The flight plan was outlined on a single sheet of paper, which was then passed through a hatchway in the wall through to a waiting communications officer. This officer annotated the codes for the various units who provided any service to the flight and then typed out the entire format onto a teleprinter and set the transmission sequence in motion. The message, known as a flight safety message was then received by all addressees within Australia and also overseas if the flight was proceeding to international destinations. Outside in the airport air terminal areas, aircraft were preparing to take on their next passengers. Trolley trains zoomed around the apron area while smartly uniformed hostesses ushered their customers onto the aircraft and the scene was a quiet dignified spectacle in those days unlike modern terrorist-ravaged operations. Everyone took great pride in being part of such a young and vibrant industry in the 1960s and 1970s and Darwin was a wonderful place to work during this period. Sometimes, the Northern coastline weather was clear and fine and at other times storm clouds threatened to ruin a smooth flight, but no-one really seemed to mind.

I began specialist training for local Darwin operations early in 1971 and a training officer was formally assigned to me. His name was Brian Wise and he wasted no time in explaining the international radio procedures that I was to use in the Darwin Flight Service Centre. His name seemed very appropriate for an instructor! My first job was to operate the International sector using High Frequency communications equipped with SELCAL. SELCAL was a selective-calling-device manufactured by Motorola and operated rather like a doorbell. Each aircraft using SELCAL was assigned a specific four-letter SELCAL code by ICAO, (International Civil Aviation Authority). The four-letter code was appended to the flight plan before departure. Units equipped with air/ground/air SELCAL received the flight plan containing the specific code and set this code on four selectors located on the SELCAL transmitter. Each console was equipped with banks of selectors and each bank comprised four black tumblers made from Bakelite. The code was set by rotating the tumblers until the desired alphabetical letter appeared in the tiny window. The aircraft identification was then marked on the front of each selector unit with chalk. The unit was equipped with a push switch, which when activated sent out a coded signal to the aircraft. The signal activated an alarm or flashing light in the cockpit of the aircraft so that the crew were immediately aware that a ground station was calling. They would simply turn the HF radio up and transmit an invitation for the ground station calling to 'go ahead," with their message. In this way, crews on long journeys did not have to listen to hours of noisy radio static in the headphones. However the system only needed to operate one way, as the ground Flight Service Officer was required to listen to the static all the time. This was because domestic internal flights were not equipped with SELCAL as the equipment was too expensive for smaller operators to afford. Several HF frequencies were used simultaneously so that variations in radio ionospheric conditions and local interference conditions did not interfere with transmissions across the entire frequency spectrum. If a 13 mHz frequency could not be used because of the ionospheric conditions, then a lower frequency could be used. Darwin Flight Service operated in the SEA3 International network, which was the southeastern Asian radio network-three area. This area used frequencies of 2987, 5673, 13288 and 17965 mHz to provide a full range of communications coverage over each 24 hour period for a range of stations including Darwin, Djakarta, Biak, Makassar, Singapore, Rangoon and even as far as Bombay. The ionospheric gases, often varied in thickness and height around the global surface and often presented some severe radio communications limitations. Variations in solar radiation caused this belt to thicken and vary its distance from the earth's surface. HF radio signals using the ionosphere to bounce radiated signals, were often distorted and radio frequencies had to be changed quickly in order to ensure continuous communications with all aircraft. The angle of bounce from the ionosphere depends upon the frequency actually being used and generally, lower frequencies would be used at night and higher frequencies would be used during the day. Therefore at 3 am you could expect to be using 2987 mHz and at 3 pm you may well be using 17695 mHz. But as with everything else in life, certain anomalies occur which cause distinct variations in this ideology. These technicalities however, are well beyond the scope of this story.

The main limitations of HF usage was its susceptibility to hash and radio static. At times this could be severe enough to attenuate the signal so badly that reception was impossible. The only alternative was to use Morse code which was less prone to this sort of interference. Unfortunately officers sitting on a seven-hour shift listening continuously to radio static, eventually suffered from long-term hearing deficiencies which often resulted in permanent ear damage. A CODAN, (Carrier Operated Device Anti Noise), was employed to reduce this problem, but could not be employed when the signal from the aircraft was very weak. This was because it reduced the sensitivity of the receiver when attempting to reduce the static noise levels. Devices like the SELCAL and CODAN were some of many attempts to reduce the static interference that prevailed in HF communications. HF was the only means of long range communication until the advent of satellite and microwave equipment. This equipment was able to convey the clearer VHF signals that were normally confined to short distances over large distances by simple means of relay. Aircraft were progressively becoming better equipped with voice-modulated radio equipment instead of using the cumbersome morse keys. Some long-range flights continued to carry Radio Officers so that the pilot could concentrate on flying the aircraft instead of being distracted by the radio. These aircraft would often convert to sending morse-encoded messages when radio conditions prevented voice signals from being received.

I found that my morse skills lacked practice and I was becoming more exposed to voice operated radio equipment. As a result my receiving speed became a little slower than it should have been and caused me some embarrassment one night while I was communicating with an RAAF C130A Hercules transport aircraft en-route from its Malaysian base at Butterworth to Darwin. The navigator was trying to relay his position to me by voice on HF radio, but interference from very poor radio conditions that night made reception of his calls impossible. I was suddenly interrupted by the navigator converting to the use of morse and sending at such a high speed, that I simply could not keep up with it. I yelled for assistance from the Watch Supervisor who was in charge of my shift. He was at his desk talking on the telephone and I was afraid that I would lose contact with the C130. He completed his call and ambled over to where I was sitting. He was a Queenslander named Bert and his rugged features and crew-cut hairstyle belied a friendly individual who was very experienced in radio and morse-code communications.

"What's the matter young feller?" he drawled.

I told him what the panic was about and with a grin, he leaned across me and scribbled the position report sent by the C130 onto my radio log. As an old operator, he heard the transmission as a second language from where he was sitting, without any effort at all. Rather humiliated, I watched him amble back to his desk and I resumed my operation with increased vigour, determined to become as good as my obvious superiors! A number of Australian aircraft operated from Darwin to Dili on the island of Timor and across the Makassar region, (now Ujung Pandang), to Balikpapan in Borneo. Routes were flown between Darwin to Ambon on the island of Seram and across to Cendrawasih on the northwestern tip of Irian Jaya. Charter pilots from various companies including the South Australia and (Northern) Territory Aerial Services (SAATAS) and the Phillips Oil Company, flew a variety of aircraft in these areas. These included the Fairchild FA24 (which was a derivative of the well known Fokker Friendship FK27 aircraft, PBY Catalina flying boats and a wide range of light aircraft like the twin engine Beechcraft Queenair, Grumman Goose Amphibian and single engine Cessnas. The remoteness of the island locations combined with the vast oceanic areas of the Flores, Banda, Timor and Arafura seas made search and rescue considerations a nightmare. For this reason, Australian registered aircraft or aircraft flown by Australian aircrew were required to report their in-flight position at regular intervals, so that search areas could be significantly reduced in the event of a mishap. Communications during the day were usually conducted on 8868 kHz with Darwin Flight Service and once the aircraft and arrived in the Indonesian controlled areas, would continue on 8820 kHz communicating with the Aeradio Officer in Makassar. HF conditions were sometimes badly affected by thunderstorm activity and pilots often communicated with each other on the shorter range VHF to obtain mutual support in the event that someone suffered an emergency and could not get through on HF to Darwin or Makassar. The restricted availability of frequencies for voice transmissions was simply dictated by the cost of the radio equipment. As a result the need for morse code operators continued well into the late 1970s.

    When I commenced duty for the first time on the Darwin Flight Service International circuit, I was keen to reach the high standards of operation set by my predecessors. The very first aircraft I spoke to on a live microphone was a Douglas DC8 airliner operated by Air New Zealand. The pilot followed normal operational procedure and made his transmission in the form of an air position report, known colloquially as an 'AIREP.' The AIREP contained the present position of the aircraft, the operating flight level or altitude, an estimated time of arrival for the next nominated position report and the current weather being experienced. The radio transmission was typical and went as follows.

"Darwin, this is New Zealand Zulu Delta, airep position on 8868."

On receipt of this message I responded with the following standard acknowledgment.

"New Zealand Zulu Delta, this is Darwin on 8868,- go ahead."

The crew would then give their position report and weather information in the standard format as follows:

"New Zealand Zulu Delta, airep position, Sandshark zero-one, flight level three five zero, estimating 150 miles Darwin at four-zero, wind three-five-zero degrees, two-eight spot, temperature minus three-five, on top broken stratiform." These details were taken down on a flight strip (see below).

NZZC	WSSS	SIP 3-12	FIR BDRY	SDK 0-34	150 DN	DN 211 ✓
DC8	1800	2123	2140	2159	2240	TAF 20/08 ✓
HJGL		11	25		2201	
WSSS/DN	350	350		350	350 ∅	

Typical flight information sector strip on the Singapore to Darwin route.

The aircraft was nearing Darwin and would have to transfer to Air Traffic Control at a range of 150 nautical miles from the aerodrome. At this point VHF radio contact could be expected and the need for further transmissions on HF discontinued. I was now required to make the following instruction to the aircraft.

"New Zealand Zulu Delta at one five zero miles Darwin call Darwin Control on one-two-five-decimal-two."

My next duty was to relay the position report to the Air Traffic Control Unit. They would then know when to expect the aircraft and also ensure that there was no possibility of collision confliction within his area of responsibility at the aircraft's current operating level of thirty five thousand feet.

Next, I typed the meteorological data on a teleprinter and sent it to a series of meteorological stations, so that they could evaluate the upper level weather conditions and variations to expected trends. When traffic conditions were heavy, a machine telegraphist would assist by operating the teleprinter for you, otherwise the job was overwhelming for one operator on his own. While the aircraft was transmitting the position report, the Flight Service officer would annotate the cardboard flight data strip with the information given by the aircraft. Usually, each aircraft was represented on a separate cardboard strip, but if the route being flown was fairly lengthy, then two strips would have to be used. The cardboard strips were placed in order of time expected for each aircraft at its next nominated position report. In this way the next aircraft call to be expected was represented by a cardboard strip at the top of the board. This ensured that no aircraft position report was ever missed and after it had reported its position, the cardboard strip would then be placed further down the board in appropriate time order. If a pilot failed to report his position, then communications checks would be performed by all stations on the international radio network, three minutes after the call was expected. If the aircraft had still failed to report after fifteen minutes from when the scheduled call was expected, then an emergency would be declared by the ground station responsible for the safety of the flight and search and rescue procedures would be initiated. The cardboard strips were each held in metal holders measuring approximately twenty centimetres long, by two and half centimetres wide. The strips remained active on the Flight Information Board (FIB) until the aircraft had landed safely or had transferred into airspace being controlled by another Flight Service or Air Traffic Control unit.

The Search and Rescue (SAR) system, was operated by the Department of Civil Aviation and employed three phases of Search and Rescue in order to indicate the level of concern felt for a missing aircraft, or an aircraft with a reported emergency condition. The Flight Service Officer or Air Traffic Controller first becoming aware of the situation, was required to declare an "uncertainty phase," and a Search and Rescue (SAR) message was then dispatched via the teleprinter system to all Air Traffic Services units along the intended flight path of the aircraft. If apprehension was felt about the aircraft safety and it failed to report by its next position report, or if other information received, indicating that there was an emergency problem, then an "alert phase," would be initiated and this automatically superceded the uncertainty phase. At the declaration of these phases the Rescue Coordination Centre was activated by the Operational Control Centre staff. If the aircraft failed to report by a reasonable time or failed to arrive at its nominated destination, the RCC staff would upgrade to a "distress phase," and a search for the aircraft would begin. Of course, a "distress phase" could be initiated immediately, if a life-threatening situation was identified by the officer first becoming aware. These situations could include, "Mayday" calls or other calls indicating that the aircraft operation was gravely impaired. Australia is a contract member of the International Civil Aviation Organisation (ICAO). One of the duties of this organisation is to standardise aeronautical procedures throughout all its members. If a search and rescue action has been commenced and is likely to involve another country, then the alerting message transmitted, must be in the format used by all ICAO members, so that there is no doubt about the problem, as soon as the message is received. The SAR system has now been updated, making extensive use of satellites and computer technology. However, before these advances in technology, flight safety messages had to be manually transmitted on a teletype network being operated by Communications or Flight Service Officers. Messages were sent all over Australia or anywhere in the world on the Aeronautical Fixed Telecommunications Network (AFTN) and this system could only be accessed by aviation or maritime authorities. The AFTN relied on a tape relay system installed between user-stations connected by either a landline or via a Radio teletype (RTT) network. Messages were transmitted using perforated paper tapes, that often had the contents of the message decoded and printed on the actual tape. However the DCA Communications Officers and Flight Service Officers were able with practice to actually read the holes in the tapes and understand the message from the patterns of perforations without the need for the message to also be printed on the tape. Each message could be transmitted to a maximum of six destinations but if you needed to send the message to more than six stations, then the message simply had to be transmitted twice. This was done by taking the tape out of the machine, typing in the next address list, inserting the tape back into the transmitter unit and pressing the "send" button on the transmitter encoder.

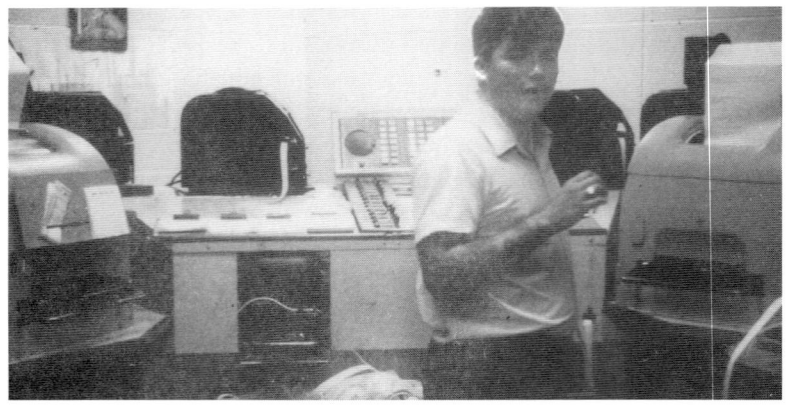

DCA Communications Officer Terry Ford on duty in the Darwin Flight Service Centre Message Switching Centre 1970 (Photo Author)

A system of priorities was employed using a coded form for differing levels of importance of a particular message that required transmission. For example, a high priority was applicable for a distress message while a low priority was used for routine administrative messages. Other classifications of messages used various priorities in between these two extremes. Each message was assigned a two-letter machine telegraphy code that consisted of two identical letters for each level of priority. This ensured that traffic was sent on time. A message annotated "FF" always related to flight safety messages of immediate concern, like flight plan and departure messages and important weather forecast and NOTAMs of immediate concern to aircraft in flight. Search and Rescue messages were coded "SS" and also included an additional code which activated a bell five times immediately it was received by the addressed station, so that the officers operating at that station would hear the message arrive and deal with it without delay. "GG" messages related to lower priority routine meteorological messages and operational messages. The lowest priority was "LL" assigned to routine administrative messages.

Communication machine telegraphy relay centres at the capital city aerodromes were always a hive of activity, especially in the early morning when the first flights began planning for the day's flying schedules. The teleprinter messages were typed out rapidly sending the flight plan details and departure details to the other units that would be involved with a particular flight. Quite often there would only be a minute of time between messages and the constant reliability of the system was a credit to the officers who manned the relay positions, a credit which has gone without notice with the passage of time. The early teleprinter machines used were manufactured by the Siemens Company and employed what was known as a "three-row," teleprinter, each equipped with an automatic tape transmitter and receiver. The "three-row," referred to the fact that there were three rows of keys on the keyboard. A later model used a "four-row" system that was not quite as well accepted by officers who had attained a high level of efficiency on the "three-row" model. The communications officers would read the coded addressees on the printer roll page copy and then tear the tape containing the same message off the tape receiver. They would then run to the teleprinter circuit on which the first addressee was located, insert the paper tape and commence sending the message. When the tape had run through to the end, the Communication's officer would then remove it and take it to the next printer to send to the next addressee. This process was repeated until the all the addressees on the message had received a copy of the required data. In order to achieve these requirements, the early communications centres were equipped with rows of teleprinters, all clattering away, sending and receiving message traffic, twenty four hours a day. Normally, two or three operators were on duty during a shift, all using the "torn tape," procedure. It was possible for one operator feeding a series of multi-gang transmitter distributor units to control up to twelve outgoing circuits. Each of these circuits in turn comprised combinations of radio circuits connected to Flight Service, Air Traffic Control,

Meteorological and Administration Units in Australia and all over the world. Notices to Airmen (NOTAMS), were the bane of everyone's lives. They were printed to advise airmen of unforseen changes in the serviceability of runways, navigation aids, or operational information, regarded as necessary for the safety of aircraft. Notam's were grouped together in a bulletin format. The bulletins were arranged in two ways, either in order of travel for long air routes or in alphabetical order for locations within a specified smaller areas. The bulletins were officially known as 'PFIBs' or Pre Flight Information Bulletins. Pilots or nominated crew members would attend the briefing office well before the flight was due to commence and obtain a two part briefing. The first was a meteorological briefing from the weather officer and the second part was an operational briefing in the form of a PFIB which concluded with the submission of a flight plan to the Air Traffic Control or Flight Service Briefing Officer. Generally, Air Traffic Control conducted the briefings at capital city offices and Flight Service Units provided the briefings elsewhere.

Pilot and co-pilot from an Ansett Airlines Fokker Friendship FK27 scheduled service to Darwin, plan the flight in the Flight Service Unit briefing office at the old DCA facility site at Alice Springs Airport on the western side of the present airport. PFIBs can be seen hanging on the wall behind the crew with pre-printed flight plan forms on the clips to the right of the picture.
Photo (Author)

PFIBs were compiled each evening and typed on teleprinters switched out of the normal transmission circuits. This was so that the paper tape transmitter could be utilised to retain or modify the information, rather like a modern day computer disc. This would alleviate the need for retyping of information each time it was requested by various users. The tape was analysed to locate the required information, fed into the appropriate tape transmitter facility and sent to the required station.

The rosters used in the Darwin Flight Service Centre in 1970 were arranged so that the afternoon shift, operating the international switching console, were relieved by the international air-ground radio operator reporting for duty on the evening shift at 5pm. When the officer on the afternoon shift was relieved, he would proceed to the operational control centre to obtain the updated PFIB, edited with all the changes for the last 24 hours. The master copies of the PFIBs for all the areas and air routes required for the following day, were taken to the Communications teleprinter relay centre where a teleprinter was isolated out of the active circuit and used to print the appropriate changes.

The updated page was copy taken from the teleprinter roll, and transferred to a duplicating machine that used a chemical paper that was used to run off the required numbers of copies likely to be demanded by aircrew of regular flights for the next 24 hours. The problem with this system was that the chemical paper was sensitive to direct sunlight and would fade if left exposed to the sun. If the routine supplies of the PFIB copies ran out during the day, the master teleprinter tapes could be used to run off more copies, except that these bulletins contained a 'stop-press' sheet which featured new information received since the last PFIBs were printed. The problem was, that approximately twenty bulletins were required for each route or area and there were approximately thirty of these areas or routes for which bulletins were routinely required! Obviously, a bulletin prepared for an aircraft proceeding from Darwin to Manila would not be much use to an aircraft proceeding from Darwin to Djakarta. The officers compiling these bulletins ensured that errors or omissions had not been made, then assemble each bulletin and return them to the operational control centre ready for distribution of the updated information to aircrew. When they had fulfilled the role of "unofficial government publishers and printers" the Flight Service and Communications officers resumed their normal communications duties. Staff usually finished the PFIB production after 8pm, when the shift officially finished, but rarely claimed overtime, such was dedication to duty.

I enjoyed the fact that I had access to information about any airfield in Australia or in the world for that matter. I often read the bulletins in my spare time simply out of interest and fascination. Aircraft had often already departed after new information was received that could affect the safety of its progress. The new information was assessed by the duty operations or briefing officer and automatically relayed by Flight Service or Air Traffic Control via the radio. The pilot could then make an appropriate decision based on the details that he had received. Pilots on charter or regular public (RPT) transport flights were required to maintain a listening watch on the nominated en-route frequency and therefore received all the necessary updated information as it became available. If the aircraft failed to respond to a call from a ground station, the operator would continue calling for fifteen minutes assisted by other network stations and aircraft. If contact was not re-established after fifteen minutes, the operator who initiated the first call would declare a search and rescue phase, which would alert the nearest airways rescue co-ordination centre (RCC) to prepare for a possible search and rescue situation. Every incident or accident was different and required a wide range of different responses. Some could be quite humorous had they not been so tragic.

One story reported quite some time ago, concerned a light aircraft that failed to report arrival at Port Keats. Port Keats airstrip is located approximately 130 miles southwest from Darwin on the coast near Joseph Bonaparte Gulf. The flight was on a charter service and had made the mandatory reports along its flight path. These reports were required so that when a report was missed, the locations where the aircraft had already passed would not have to be searched. Flight Service called the missing aircraft for a scheduled arrival report but could not re-establish contact from the time the aircraft was last heard. The authorities in Darwin were notified at the required time and they contacted the local authorities near Ports Keats. A police officer was asked to go and check the airstrip to see if the aircraft was there. Maybe the pilot had simply suffered a radio failure. The Darwin authorities were startled when the policeman declined to go because he already had an urgent job to attend. Darwin argued the importance of their needs and the beleaguered policeman was ordered to take his four-wheel drive vehicle out to the airstrip and report back before attending to anything else. Driving in the bush is not quite as easy as the some city-based folk would imagine. Hence sometime later, the officer telephoned back to report that there was no aircraft at the airstrip. The Darwin authorities thanked him for his trouble and advised that an air search would be ordered immediately for the missing aircraft. Before they terminated the call they tried to be sociable and inquired what the other job was that was so urgent. They were stunned by the weary reply. He was investigating a report from one of the local aboriginals, who thought that they had seen an aircraft crash just down the road!

It was during this period that I managed to find the time to join the Darwin Aero Club and recommence my flying lessons. The club was very busy and held numerous social nights when we would sit around the Barbecue and discuss the virtues of slide-graphic navigational computers and the circular style computer. Both were simply mechanical slide-rules that were available in large or small sizes. The smaller ones were favoured because you could usually fit them into a shirt pocket for use while you were flying. The slide-rule was simply used to compute aircraft navigation tracks and headings, with various wind direction and velocity inputs and also groundspeed and time intervals for flight planning purposes. They were not complex and even used by weekend private pilots as well as commercial pilots. The club Manager was Peter Denholm and the Chief Flying Instructor was Warren Gengos. Peggy Hassell controlled the office, while instructors Len Spanton, Tom Sparks and Don Mitchell worked hour by hour with a large number of student pilots like me. Peggy still lives in Darwin working for the Palmerston City Council.

Peter flew with me on my pre-solo check ride in the club two-seat Victa 115 trainer, registered VH-APV and on the 8 December 1970, I flew my first solo from runway 36 at Darwin Airport. The monsoonal wet-season had begun and I was very hot and sticky from the weather as well as from fierce concentration on what I was doing with the aircraft. I remember Peter hopping out of the Victa, then leaning in to fasten his now vacated seat-harness and shouting over the engine for me to "do one circuit and then come in for a debriefing." I hardly remember the flight except for when I turned onto the base leg of the circuit in the correct position at ninety degrees to the runway final approach path and was cleared to land by the control tower. I glanced sideways at the empty seat beside me and then back at the runway. As I lowered the flaps and reduced power it began to rain and I remember watching the droplets run off the windshield in the slipstream. I landed safely and taxied back in to the Aero Club apron feeling naturally exhilarated. Someone must have told the Control Tower that this was my first solo, because I remember them clearing me to backtrack on the runway and also congratulating me, which put a nice touch to the whole exercise. I parked the aircraft and walked back to the clubhouse where several on-lookers clapped as I walked in and I was again congratulated. It was a great day and I was starting to like Darwin very much. As the next few weeks went by, I gradually began to build up my flying experience and spent many hours flying to and from the designated flying training area located on the Cox Peninsula. The Delissaville airstrip was available for us for training and we practiced engine failures, glide approaches and short field landings on a regular basis there. On one occasion, on the return flight from Delissaville, I was entering the Darwin Airport circuit pattern, when an RAAF C47 (DC3) transport aircraft arrived with an undercarriage problem. I was asked by the control tower to fly behind the aircraft to see if the main undercarriage could be seen properly down from my airborne perspective. I was a bit surprised at this request because

I was only a student pilot but had a look anyway. I could not be fully sure but as I peered up at the transporter, everything seemed OK. When I first saw it fly past me, I could not help but think again about the Kyeema which was similar, as I watched the aircraft lumber around the circuit and land safely.

Piper Cherokee 140 VH-PDP parked outside the Darwin Aero Club Nov 1970. The clubhouse was destroyed during Cyclone Tracy 4 years later. (Author)

After my first solo on 8 December 1970 in a Victa 115 VH-APV off Runway 36 at Darwin International Airport (Author)

Before the onslaught of Cyclone Tracy and the death and destruction that it wreaked on the city of Darwin, the area was characterised by sparse vegetation with wide-open spaces available for recreation. People could often be seen having picnics or barbecues and there was a feeling of freedom and friendship that was unique at that time. I was very interested in the history of Darwin especially through the Second World War period. There were still a lot of historical places to see including a lot of wartime airstrips along the Stuart Highway to the south of Darwin. I carefully studied the routes flown by the Japanese when Darwin was bombed and used to fly the Victa around the same flight paths. I flew them as close as I was authorised to do so, simply so that I could gain an airborne perspective of what the area must have looked like to the enemy bomber pilots and the defending allied pilots at that time.

Darwin city taken from the Victa 115 at 1500feet, looking southwest during December 1970. The Japanese attack on the city came from the area near the top left of the photograph, flying from left to right towards the harbour. The old Darwin hospital can be seen in the middle lower picture. The esplanade is devoid of the current high-rise city buildings and Mindil Beach (site of the current Casino) and Botanical Gardens area are still wide-open spaces. (Author)

From a war historian's point of view, Darwin is a very interesting place to visit. There are many relics of this era and although the focus is rightly on the men of the military who defended the area before, during and after the Japanese attacks in the February wet season of 1942, the men of the Department of Civil Aviation also played an important role. On 19 February 1942, DCA officers stationed at the Civil Aerodrome in Parap were manning the operations building. They spotted a dogfight between three aircraft to the north of Darwin and realised that an attack had commenced. Radio operators Bruce Acland and Ted Betts ran to the radio room and sent out an encoded message that Darwin was under attack. The historic telegram was sent via the DCA Aeradio station at Cloncurry and read as follows:

> DARWIN BOMBED MACHINE GUNNED 20 JAPS 30XMNS 1000 NINETEEN CIVIL BUILDINGS AND RADIO ROOM MINOR DAMAGE FIRE TENDER HANGAR SOME RAAF BS7GS BURNT DROME O.K. THINK ELEVEN MILE O.K. ALL STAFF O.K. STOP LATER. SECOND ATTACK STARTED 1210 STILL IN PROGRESS
>
> BETTS

He had the foresight to lock away all DCA documents before cutting the power to the transmitter and rushing to the nearest air-raid trench. Bombs were already falling as he reached the trench and after this raid, he and the others found a number of buildings including the DCA administration building on fire. Nearby an ammunition dump had been hit and bullets were still exploding. They were able to salvage the DCA transmitters and receivers but could not operate them because the DCA power plant had been badly damaged and the town mains electrical supply had been cut.

Communications were cleverly restored using a direct link to the Coastal Radio Station and for days, DCA staff worked around the clock to restore all DCA communications facilities in the Darwin area. During the bombing of Darwin, extensive use was made of the DCA sea-launch directed by DCA officer John Waldie, who tirelessly plucked drowning sailors from the water during the bombing, an action for which he was awarded the MBE. The DCA Aeradio station at Darwin was withdrawn fairly soon down the track and relocated inland to Daly Waters after the first air raid and was replaced with a temporary station. At this time the Aeradio stations were equipped with the standard communications inventory including a High Frequency (HF) version of the RAAF Adcock system used to obtain Direction Finding (DF) bearings from aircraft in flight. DCA adopted a system using the 2-12 mHz frequency range and installed these units at many stations including Darwin and Groote Eylandt. Although the units were temporarily made available for RAAF use during the war, they were still available for use with civil aircraft. At the time of my posting to Darwin, the Darwin Flight Service Centre supervised a network of Flight Service Units located at Alice Springs, Tennant Creek, Katherine and Gove. The Daly Waters Flight Service Unit had also come under the jurisdiction of the Darwin radio network supervisors but had closed down its operations before I arrived in the NT. A viewpoint was being generated nationally by people who wanted to emulate the systems used in the UK and USA without consideration to the fact that Australia does not have the same population base to support such ideologies. The result has been the passing of the Flight Service system on the flimsy basis that it was not cost effective. Whatever the new views of aviation administrators are today, there is no doubt that the remote Flight Service Units provided an operational service that demonstrated the highest standards of excellence to all aircraft.

The privatised replacements of this service working at a somewhat lower operating standard is proving to be quite an expensive outlay for local aerodrome operators. Sadly, Darwin Flight Service finally ceased operations in the early 1990s after the closure of all its sub-units in the Northern Territory and succumbed to the advances of new ideas and technologies. The Operations Building that had housed the Flight Service Centre, the Communications Message Switching Centre, the Air Traffic Control Operational Control Centre and the Bureau of Meteorology Briefing and Forecasting office was demolished without trace by the RAAF in the later part of 2000 thus ending an era of Northern Territory Aviation.

Doug Whitfield

World War 2 airstrips in the Darwin NT area

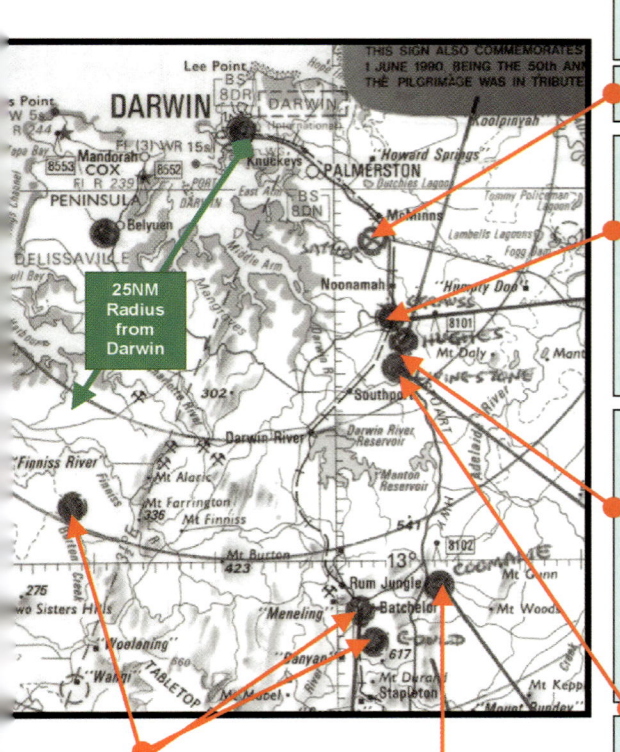

Doctors Gully Darwin used by 43 Squadron RAAF Catalina's from 9 April 1944

MKT Civil Airstrip near RAAF Sattler

RAAF Strauss initially constructed by US 808th engineer battalion (aviation). Later works by allied works council & No1 Airfield Construction Squadron RAAF. Known as the 27 mile and later named after Captain Allison W Strauss CO 8th Fighter Sqn, 49th Fighter Group USAAF, killed over Darwin 27 April 1942. Used by 49th Fighter Group (P40 Kittyhawks), 452 Sqn RAAF (Spitfire Mark Vc & VIII) and no 549 RAF, (Spitfire Mk VIIIs).

RAAF Hughes constructed by US 808th Engineer Battalion (aviation) and 1 & 9 Airfield Construction Sqns RAAF, named after the Department of Mines, who recommended the site in early 1942. Used by No 1 Photographic Reconnaissance Unit RAAF (Buffalo, Lancer, Wirraway, Lighting aircraft). No 54 Operational Base Unit RAAF, No 13 Sqn RAAF (Hudson), 34 Sqn RAAF (Dragon, Tiger Moth, Anson) and No 2 Sqn RAAF (Hudson. Beaufort and Mitchell Aircraft).

RAAF Livingstone, initially constructed by US 808th Engineer Battalion (aviation), No 1 Airfield Construction Squadron RAAF and the Allied Works Council. Known as the 34 mile strip & later named after 2nd Lieutenant John Livingstone of 9th Pursuit Sqn, 49th Pursuit Group USAAF, killed here in a crash landing on 4 April 1942. Used by 9th and HQ Sqns 49th Fighter Group USAAF (P40 Warhawk) No 77 Sqn RAAF (P40 Kittyhawk) No 457 Sqn RAAF (Spitfire Vc and VIII), No 54 Sqn RAF (Spitfire Mk VIII) and No 548 RAF Sqn (Spitfire Mk VIII)

Other RAAF airstrips in the Darwin area may be visited subject to landowners approval. These include Batchelor famous for its radio signals units and a visit by General Douglas McArthur, RAAF Gould, RAAF Blyth and RAAF Pel. Aircraft using these strips include B17 Fortresses, Wirraway, Vengeance, A24 Dauntless, P40 Warhawks & Mitchells.

RAAF Coomalie constructed by No 1, 9 & 14 ACS RAAF. Used by No 1 Photographic Reconnaissance Unit RAAF later designated No 87 PR Sqn RAAF (Lighting, Wirraway & Mosquito) and No 31 Sqn RAAF (Beaufighters). The airstrip was a vital Intelligence centre for HQ NW Area. No 1 Medical Receiving Stn RAAF was co-located here

25NM Radius from Darwin

Gove on the Gulf

In July of 1971, the Flight Service Centre Supervisor Eric Bergholtz called me over to his office for "a quick chat." He told me that a new Flight Service Unit was being opened at Gove on the east coast of the Northern Territory. This location faces the mighty Gulf of Carpentaria and consisted of a few locations inhabited by the Yolngu clans and other Aboriginal groups. Eric needed a single staff member urgently for a temporary period, until a more permanent married officer could be found. To this day, I am not sure why a married officer was required, because the accommodation of the day was more suitable for single officers. But DCA policy dictated the terms and I, the eternal humble servant, did as I was asked and cheerfully accepted the task. The expected duration of the transfer was approximately three months.

The new Flight Service Unit was needed to provide an in-flight traffic information and search and rescue alerting service for the substantial increase of air traffic into the area. An air tragedy at Gove had taken place recently with fatal consequences for all on board and this added momentum to the need for a radio unit to support air traffic in such a remote area. The ghosts of the Kyeema seemed to following me. Everything seemed to be happening because a crash had occurred. I was keen to play my role in providing the mechanisms that would afford air travellers the maximum protection possible, especially in the Australian outback. An Air Traffic Control Tower had even been briefly considered for the airport, but the expense was not considered to be justified. I had never seen this part of the Northern Territory or the Gulf of Carpentaria and as this sounded like another adventure beginning I was very pleased to volunteer my services!

On the day of departure, I flew out of Darwin on an Ansett Airlines FK28 Fokker Fellowship jet proceeding via Groote Eylandt. Groote, as it was known locally is a rather large island bearing manganese deposits amongst other things, on the western side of the Gulf. We landed there and remained for an hour or so before setting off again for my new temporary home at Gove.

Gove airport is built on the site of an old RAAF wartime airfield. Evidence of the RAAF wartime operations there was still clearly visible. The Gove Peninsula was named during WWII after Pilot Officer William Julius Henderson Gove who was a Navigator on an RAAF Lockheed Hudson bomber. Gove was killed on 20 April 1943 after taking-off from Milingimbi on the northern Arnhemland coastline of the Northern Territory. Not long after the crash, a new RAAF airstrip was constructed on the eastern tip of Arnhemland and was named Gove in honour of the fallen airman.

The Nhulunbuy peninsular was occupied by Aborigines, many of whom lived at the Yirrkala Settlement in the early 1970s. The settlement was located on the edge of the Nabalco mining operations that had been permitted to dig up the very rich bauxite deposits that had been discovered there. The Yirrkala Aboriginal mission was operated by the Methodist Church and is an Aboriginal term for "running water" or similar meaning. This is probably not surprising since the Yirrkala mission is actually located on a free-running creek. The mission was a prohibited site for other than mission workers and their Aboriginal Parishioners, however entry could be made by special invitation or permit for specific purposes. I was fortunate indeed to be permitted to explore the area for a day and found the local inhabitants very

Headstone of Pilot Officer Gove located at the Adelaide River War cemetery south of Darwin NT
(Photo Author)

industriously manufacturing Aboriginal artefacts for the Darwin tourism industry including boomerangs, didgeridoos, and a variety of hunting weapons. As I crossed the freshwater creek, I passed Aboriginal women standing in the estuary washing their clothes and bathing their babies and toddlers. It was a very peaceful scene in this natural tropical paradise.

The Nabalco Company had commenced mining operations and raw bauxite was literally being blown out of large open cut mines in huge proportions. It was transported in its raw state on conveyer belts covering over 27 kilometres to the seaport. One of the continuous conveyer belts was one of the longest in the world and measured 18 kilometres long. The bauxite was stockpiled at Dundas Point and loaded onto ore carriers bound for Japanese and European destinations. I privately hoped that the local environment would not be destroyed and that the natural beauty that existed would somehow be preserved. I need not have feared much because the Nabalco company had already assessed how they were going to return the area to its natural state when operations were completed site by site. When I arrived at Gove Airport, I discovered that there were no roads in or out of the township. It was totally isolated from any other form of civilisation except for very rough bush tracks. 800 construction workers had already moved in and were busily building the various mining plants and a township for the families who were expected to move in at a later date. The township was close to completion and featured a brand new hospital, a secondary school, supermarkets and air-conditioned homes. I lived on the Gove peninsula before the town was inhabited as it sat there in the tropical sun like El Dorado in the Amazon jungle or like a science fiction movie where the population had vanished. This impression was somewhat flawed by the various construction vehicles still driving around performing the finishing touches to these facilities, but the experience for me was unique just the same. The whole area was located inside the aboriginal reserve and the mining company imposed very strict regulations about the behaviour of its personnel with respect to the aboriginal people.

Doug Whitfield

I was met at the airport by the newly appointed airfield Officer in Charge, Adrian Shee. He was very enthusiastic about the role of the Flight Service Unit in this newly developing area and briefly explained my duties and accommodation arrangements. Arrangements had been made for me to camp at the old Europa Launch and Development Organisation (ELDO) site which was located some fifteen kilometres down a graded dirt track to the south of the peninsula. Adrian presented me with a rather dilapidated Ford Falcon station sedan painted in the usual Department of Civil Aviation yellow colour with the large white wings insignia painted on the front doors. Everything in the Gove area was contaminated with red dust and this car was no exception. Vehicles like this were transported from Darwin on old landing barges and could be used quite easily on the well-worn tracks between the airport and the work sites. Adrian warned me to be careful while driving because the surface was loose and rather like driving on loose ball bearings. He advised me not to travel above 20 miles per hour until I was used to the slippery surface on the peninsular. Being a typical 21-year-old, I failed to heed his advice. I set off enthusiastically down the track at high speed, eager to investigate my accommodation. A thick cloud of dust billowed behind me as I sped along through the tropical bush. Up ahead, the track curved quite sharply away to the south and as I approached the bend I began to steer around it. The car did not turn and as much as I madly turned the steering wheel, it continued to slide on the loose surface, in a straight line ahead! I knew I could not take the bend and jammed the brakes full on, which worsened the situation and I entered an uncontrolled skid. My initial horror turned to outright panic as the Falcon left the track at high speed, careered through the bush and crashed into a fallen log. Luckily, I had put my seat belt on and suffered only a few minor cuts and a good shakeup. Feeling very foolish and humiliated, I coaxed the wagon back to track in a series of jerks and jumps. My posting had not started off very well and my new boss viewed the damaged vehicle somewhat gravely. I needed no further encouragement to drive at the recommended 20 miles per hour and spent the rest of my time in Gove driving very sedately indeed. My accommodation at the ELDO site had been used for tracking rockets from the Woomera range in South Australia in the 1960s. The rocket testing had long since ceased, but the site accommodation at the remote Gove location was still being used for Commonwealth Government workers operating in the area and operated in the same manner as the Commonwealth hostels in Darwin. All accommodation in Gove was spartan and consisted of small portable metal sleeping rooms which each one containing a single metal bed, a metal wardrobe, small table and a chair. These were interconnected by wooden catwalks chemically treated to combat attacks from white ants and other bush borers. The catwalks also led to the canteen, shower blocks, games room and bar. Unfortunately, the catwalks were lit at night and attracted a large variety of very large bush spiders. An early morning encounter with their very strong webs was certainly not for the faint hearted! On one occasion, very early in the morning, I was having a shower before going out to open the airport for the day when a movement caught my eye. There on the shower curtain was the largest Red-Back spider I have ever seen. Even with my personal fear exaggerating its size, it was still a huge specimen. I gingerly stepped out of the shower and grabbed a shovel that was leaning against the wall. I have no idea why a shovel had been placed in the shower block, but it proved useful as I swept it in a wide arc around me and smashed it against the shower curtain so that the spider was squashed between the curtain and the shower wall. Unfortunately, I also ripped the curtain and put a nasty indentation in the wall but I had achieved the aim. I was well and truly awake which was just as well as the bush residents had not finished their harassment of my morning progress. As I stepped out the shower a King Brown snake reared up and decided to demonstrate its prowess at attacking humans and I was obliged to use the shovel again in order to make good my progress back to my room to get dressed for work! I was attacked again at the airport by another snake on the same morning but that is the nature of the duty in remote locations and is certainly another story.

The vacated Europa Launch and Development Organisation (ELDO) temporary accommodation blocks taken over by the Commonwealth Government for employees working on the Gove Peninsular (1971) (Author)

Electrical Power to the site was provided by a remote powerhouse equipped with four huge diesel generators. The engineer responsible for the powerhouse was an interesting character called Leo who looked more like a university professor than an engineer, but he was very good at keeping things running just the same. He had a passion for classical music and was a keen high fidelity (HiFi) stereo recording fan. His collection of records in such a remote area was quite impressive and he often played his stereo very loudly blaring the intricacies of Beethoven and Mozart, the melodies of which were alien sounds indeed for his bush listeners. Nearby, a small above-ground swimming pool had been erected for use by dirty, dusty workers to lounge in, during the evening, while cooling off from the hot tropical weather. On each morning shift, I rose at 5am, battled the bush spider webs on the way to the shower block and then had tropical breakfast. After driving to the airstrip, I inspected the runway for potholes and for sleeping animals and checked all the airport facilities. I often found local natives lying about on the warm bitumen and they proved very reluctant to comply when I respectfully requested

Author outside the Gove Flight Service Unit 1971 (Author)

Gove Flight Service Unit today converted into a passenger air Terminal (Author 2001)

that they move away from the dangers of landing aircraft. As the sun rose in the east, the bush began an amazing transformation from the rich reds of the sky, to a deep blue and gradual lightening of daybreak to show the magnificence of the gum trees and the rainforest. The next duties required completion of a weather report, opening the radio room and calling Darwin on the Single Side Band (SSB) link, to advise them of the weather and that the station was open for business. Darwin then relayed all the flight plans for aircraft that would be proceeding through the Gove FSU area of responsibility. They also sent operational data they had received overnight that would affect my aircraft operations.

When the Gove Flight Service Unit first became operational in the 1970's, it was responsible for an area within 30 nautical miles radius from the airport from ground level up to an altitude of 5000 feet. The flight plan information received from Darwin was converted to data written on cardboard flight watch strips that were placed on the Flight Information Board on the radio console. As soon as the aircraft shown on these cardboard strips departed, the departure time was recorded with details of scheduled times at reporting points, destinations and any new operational information relayed to the aircraft. If several strips were active simultaneously, an analysis was made of all the flight paths, so that if a collision risk existed, the pilots could be notified in good

Gove Flight Service Unit Communications Consolette 1971. Note the glow of the CR6-B HF receiver (below left) (Author)

time to take avoiding action. The Flight Service Unit was equipped with a standard AWA CR6-B HF radio receiver that was used to monitor all the HF bands from 200kHz to 25mHz. There was not much to do at night after the station had closed down and I often remained behind to use the receiver and listen to a variety of domestic and international radio programmes. I sat for hours tuning up and down the radio bands listening to international aircraft and radio stations. The drive back to the ELDO site was lonely and full of tropical sounds. Sometimes, I stopped the car and looked up at the night sky which was nearly always crystal clear. I soon became interested in night navigation and easily found Orion and the Southern Cross constellations. Satellites and shooting stars were often visible and I sat and watched them until the last vestiges had disappeared from view. I used to wonder if any non-earthly beings were up there communicating on frequencies that I could use at work. After a while I headed back to the ELDO site where the boys could be found at the bar, all somewhat inebriated, but pretty happy to share a last drink with me before we all 'turned in' for the night. I was a comparative newcomer to the outback environment and readily absorbed with interest, all I could find out about my surroundings. I owned a second hand German made Voightlander single lens reflex camera and spent many hours trying to photograph the many unusual things around me. One night while off duty, I sat on one of the catwalks, fiddling with my camera. It was already quite dark when I became aware of a sound like an Aboriginal didgeridoo musical instrument playing nearby. If an Aboriginal corroboree was in progress, I would like to visit them and take a photograph. I couldn't pluck up the courage to go alone and went off to the bar to muster some help. The men drinking there, left me in no doubt that they were not interested in trooping off into the bush to look at a corroboree and reprimanded me with, "...if there were any natives here they certainly would not be playing "didgeri-bloodydoos" at this time of night!" I was not convinced and insisted that they at least come out and listen with me. After a short argument about the matter, they decided that I should be

70

given a fair hearing and grabbing an extra stubby of beer each, they trooped outside to where I was standing. I became very excited, as I could still hear the noise, but could not understand why my comrades stood looking puzzled, vowing that they could not hear anything like a didgeridoo. I insisted again and again that they listen harder and they stood with their heads cocked on one side, listening as though their lives depended on it.

One of the locals called Davey, had been in the bush for over 25 years and told everyone that "...if anyone knew the sound of a didgeridoo, then he would, especially since he had recently learned to play one." This was met with some scepticism, until he suddenly stood erect and said, "you mean that noise...there?" He waved his hand up and down in sympathy with the rhythm that I had been listening to for the whole time that we had been outside. I was ecstatic! The others however, still looked very puzzled. Davey suddenly set off in the direction of the noise declaring that he would take us straight there. Nervously, I fingered my camera as we trudged by torchlight through the bush. The remainder of the group followed along behind, quietly cursing and complaining that they were running out of beer. The sound grew nearer and nearer, so much so that a previously unconvinced member muttered, "geez, I didn't know there were any Aborigines out here, thetas fer sure." We continued to march through the undergrowth for another ten minutes before breaking out into a cleared moonlit area. In the centre of the area was a rickety old shed. Davey marched up to the front door and pulled it open in the torchlight. We were all overwhelmed by the noise that I had mistaken for a didgeridoo that was now emanating from an old well-worn water pump grinding away inside the shed. Davey performed a little dance and gleefully roared, "there's yer didgeri-bloodydoo mate!" and threw his head back, laughing himself into insensibility. He ran up to where I was standing, slapped me on the back and bellowed, "...an' you, yer silly bugger, owe us all a beer!" He waved a hand in the direction of the group all standing around and they immediately applauded his suggestion. I humbly admitted my mistake and obliged. Though small and isolated,

Gove was an interesting place in those early days. Most of the men who worked here had lived in the territory all their lives and spent many hours in the evening talking about days gone-by. I listened eagerly to their stories and enjoyed their good outback humour. I grew to respect these people for their no-nonsense attitudes. They were the heart of this land and I began to feel very sorry for the city dwellers. Life could be quite varied on the Gove peninsular and our little airstrip was rapidly turning into a very busy airport. We still did not have runway lighting, although they had been scheduled as a project for the not too distant future. Runway lighting was generally only needed for the Flying Doctor aircraft but there were no RFDS aircraft in the NT and so this service was performed by the Northern Territory Medical Service who operated twin engine De Havilland DH104 Dove aircraft. The pilots were mostly hired on contract from Trans-Australia Airlines for tours of duty so that they could accrue experience as pilots in command on multi-engine aircraft. A number of medical evacuations were required from Gove while I was there, mainly as a result of industrial accidents or misbehaviour. The Gove Police normally contacted Adrian and if he was unable to attend himself, he called out one of the other staff members. Whilst we had a primitive telephone system around the local area, there was

Gove Emergency runway lighting flare trailer in 1971. The flares were fuelled by kerosene. (Photo Author)

no telephone link between Gove and Darwin. As a result, Darwin had to be contacted either by using our SSB radio or the Police HF radio to alert someone to task an aircraft to fly into Gove. We immediately went down to the airstrip and hitched the flare trolley onto the back of the station sedan. Flares were dropped every 300 feet or so along each side of the runway. The flares consisted of a metal kerosene filled base with a wick assembly on the top. When the wick was lit, a three-inch flame was provided that could be seen by the pilot as he circled, preparing to land. We normally placed all the unlit flares from the trolley onto marked points on the runway edge on each side. We then drove down the airstrip, leaning out of the car with a lighted taper and brushed over each wick igniting them as you went. The pilot was then informed by VHF radio of the lighting pattern being used so that he could properly and safely plan the approach to land. After he had departed again, we would simply drive back down the airstrip leaning out of the car to replace the brass tops back on the flares to extinguish them. When they had cooled down we picked them up and returned them to the trolley and storage. I had little to do when I was off duty and as a keep fit exercise, began to assist the airport refueller move 44-gallon fuel drums by hand around the airport. Gove was being visited by numerous business jets as well as the new Ansett FK28 jet service and the need for refuelling was increasing dramatically. Unfortunately, the only way fuel could be transported, was by 44-gallon drums from Darwin via the landing barge service. An old battered semi-trailer truck transported the drums from the wharf and conveyed them down the track to the airstrip. After unloading, the drums were rolled to the top of the underground tank inlet and the contents poured in by hand. The pipe attachments for refuelling the jets only fitted the underground tank and could not be used on the 44-gallon drums. This was mainly for safety reasons to prevent possible contamination of the fuel and also because of the potential fire hazard. It was quite demoralising for all concerned to spend the whole day rolling heavy fuel drums up and down the mound from the semi-trailer, only to have one jet aircraft taxi in and guzzle the labours of several hours work in the space of fifteen minutes! The process was repeated every day to refill the underground tank. I did not have a lot to do while I was off duty and so I used to help the local refueller, Ron Gollan to roll the drums up to the top of the mound. It seemed a very good way to keep fit. On a daily basis we battled red-backed (widowmaker) spiders that seemed to think that paradise was making a web on the bottom of a drum and our operation was therefore quite perilous! One of the aircraft that visited on a regular basis was an executive jet operated by the mining company. Though small, it was crewed by two pilots and afforded the luxury of a cabin attendant to nurse the needs of the visiting executives. The company did not carry refreshments already used and so if a plate of sandwiches or cheese had been unwrapped but not used, the attendant used to throw it in the waste bin on arrival at Gove. The same went for opened bottles of liquor that had not been fully used. One day, Ron jokingly told her that he and his assistants could "do with a feed" and from that day onwards she used to bring over the left-over food and bottles of opened liquor for him. I often used to sit on the mound with him after rolling a few drums and view the goods from the aircraft with some relish. Executives certainly had a good life. The sandwiches were great and so was the cheese and sweet cakes. We lined the little bottles up along the concrete and made selections well into the sunset. It was a lot of fun and we went back to the settlement very happy!

As the open cut bauxite mine became more active, large chunks of the peninsula was literally blown up and transported along the conveyer belts to the stockpile at the wharf ready for export. The explosions became bigger and bigger until they became a hazard to overflying aircraft landing at the airfield. A procedure was developed where detonation was not effected until a clearance was received from the Flight Service Unit, advising that no aircraft were in the area. I used to go and watch the operation from a discreet distance. The mine crew prepared the explosives in a herring bone pattern then pre-set the fuses and called the FSU for a clearance, "to blow." When the clearance was given, the crew drove their four-wheel drive vehicle into position and left the vehicle with motor running. The engineer lit the delayed action fuse, jumped back into the vehicle and made a rapid exit from the area. A few minutes later the entire ground area erupted, blowing hundreds of tons of raw material into the air. When the dust had settled, the airport was notified that the operation was complete and huge earth-movers began the removal process. There were many colourful and memorable people of all nationalities working on the site and the area hustled and bustled relentlessly for twenty-four hours a day.

We sometimes spent our leisure time trudging around the bush looking for old wartime aircraft and other relics from the occupation of allied forces on the peninsula during WWII. The wreckage's of an old Ventura and a Beaufort bomber from the RAAF lay near the airstrip. I was amazed by the fact that after all the years of lying exposed to annual variations in seasonal weather and also bush fire activity, that the painted serial numbers could clearly be seen. Information from the RAAF in Canberra outlined what had happened to each aircraft, both of which had been written off in accidents and stripped for spares. The old military seaplane concrete moorings could also still be seen on the opposite side of the bay from Dundas Point although access was difficult unless you had access to a boat.

Wreckage of an RAAF Ventura bomber still on the edge of the airstrip in 1971. (Photo Author)

Another interesting feature of Gove was the intense activities of No 321 Radar RAAF and a wide range of other RAAF units including 13 Squadron, 83 Squadron, 42 Squadron, 56OBU and 39ZFC that operated during the war years. The radar site is now the carpark of the Yirrkala School!

Remains of an RAAF Bristol Beaufort bomber near the Gove airstrip 1971. The aircraft serial number A9-501 was still visible after numerous bush fires and extreme weather conditions. (Author)

The power supplies from the mine site main powerhouse provided the power for the airport and quite often failed. Huge batteries installed behind the radio room automatically switched on when the mains power failed and waited until a mobile diesel power unit out on the airfield activated. Once the mobile diesel unit activated, the battery units switched off to conserve power but continued to charge ready for the next supply failure. This ensured continuous operation of the radio and navigation facilities. On some occasions the mobile power unit failed to start automatically and had to be manually activated. If the unit failed to take over from the batteries, a safety delay of fifteen minutes was applied before a manual start was attempted. This was to ensure that the motor didn't start up on its own, while a manual start was being attempted and injure the operator. The old unit was quite belligerent and often refused to start manually with the operation becoming quite strenuous. Meanwhile the officer on duty in the Flight Service Unit monitored the power controls to verify that the unit had taken over from the mains power supply. However, one morning, the power unit failed while I was on duty and I alerted Adrian by intercom to come down to the airport and start the motor manually. Some time passed and the monitor light did not illuminate. I looked out on the airfield but could not see Adrian or his car. As time went on, I became quite concerned about the drain on the huge marine batteries and began to worry that the station would go off the air altogether. I tried to contact him on the intercom but no-one knew where he was.

Finally, I decided to try the diesel again and activated the starter button located on the Flight Service radio consolette. Happily the engine burst into life, relieving the battery units until the mine mains power could be returned to service. Without warning, the door of the radio room burst open and Adrian staggered in, covered in soot and oil. This was an unusual sight as he was normally impeccably groomed even in the bush! He was literally sitting astride the diesel trying to manually start it and was about to succeed when I hit the button. It gave him a rude shock as it prematurely burst into life showering him in exhaust fall-out! Despite the grievous circumstances I could hardly suppress my mirth at having just blown up my boss!

Eventually, my three months came to an end and a married officer was assigned to the station from Darwin. Soon the township would be receiving its first influx of families who would be doubtless overawed at the remoteness of their new home. As I looked down from the comfort of the FK28 jet along the northern coast of Australia back to Darwin, I suddenly realised just how remote these outback places were. With a small sense of pride, I was beginning to feel that I had contributed a little, towards the development that was taking place here.

A Tale from the Track

As an unmarried officer, I was soon in great demand again to provide temporary staff relief at other Flight Service Units in the region. The South Australia/Northern Territory region was controlled by two main centres at Darwin and Adelaide. These two centres controlled a network of Flight Service Units located at Gove, Katherine, Tennant Creek, Alice Springs, Leigh Creek, Whyalla, Ceduna, Mount Gambier and Broken Hill. Although Broken Hill was located in NSW, the unit was administratively easier to handle from the Adelaide Flight Service Centre instead of Sydney and so it formed part of the South Australia/Northern Territory radio network.

I had already been on temporary transfers to some of these stations and held the appropriate qualifications, so I could freely be assigned to any of them to fill in for staff deficiencies until a permanent transferee could be found. A call for me to pay a visit to the Flight Service Centre Supervisor in Darwin did not therefore come as a surprise. Usually it meant that I was about to go on another visit somewhere. I knew that a number of officers around the stations had been taken sick or about to go on annual recreational leave and temporary relieving officers had to be found to avoid build ups of overtime funding.

The supervisor at Darwin was Eric Bergholtz who always struck me as being a very fair and rather humane sort of character. I found him sitting in his office studying a pile of radio technical papers. Looking up at me, he simply smiled and said, "Alice Springs?" We both knew the form and I nodded in agreement. I simply asked, "when?" He told me that the wife of one of the officers "down at the Alice," had been taken seriously ill as a result of a difficult pregnancy and had been flown to Adelaide for urgent specialist treatment. Eric leaned back in his chair, removed his black framed spectacles and apologised for continually disrupting my life with temporary transfers, but I assured him that I didn't really mind. It was all experience and a bit of an adventure.

I decided to drive down to Alice Springs, so that I could have the benefit of my own vehicle on arrival there. The car was a newly acquired light-blue coloured MGB sports car but hardly the sort of vehicle for driving around in the bush! But I decided that it was probably unwise to leave it in Darwin with no one to look after it properly. I packed up my things at the Ross Smith hostel where I was staying and readied myself for the three-day drive down the Stuart Highway. This drive would take me through Northern Territory panoramas including some well-known World War II airstrips located along the way. Australian air force fighter aces like Bluey Truscott and Les Jackson came to mind as I hurtled past points of history like the Strauss airstrip where Number 76 Squadron RAAF, operated 24 Curtiss Kittyhawk fighter aircraft. A few kilometres down the highway I thought of the ghosts of those who had served at the Hughes airstrip, where number 13 squadron RAAF operated 18 Lockheed Hudson twin engine bombers. Further on, was the camp where No 77 squadron RAAF who operated another 24 Curtiss Kittyhawk fighters. Other airstrips passed me by at Coomalie Creek, Batchelor, Gould, Pell, Long, Fenton, McDonald and Pine Creek. I thought about what it must have been like camped there flying aircraft types like the Bristol Beaufighter, CAC Wirraway, Brewster Buffalo, Lockheed Lightning, and Republic Lancer that were operated by American and Australian airmen living there during the wartime period. I paid a quick visit to the Adelaide River War Cemetery to pay my respects and I was soon on my way again, heading off down 'the bitumen," as the Stuart Highway is known, en-route to Katherine. I already have a few memories of this place! The Civil Radio Officers who worked for AWA and DCA keeping track of both civil and military aircraft certainly did not have an easy life living in these very harsh conditions with no air-conditioning.

After four hours on the very narrow strip of bitumen that was the Stuart Highway I entered the main street at the Katherine Township. I found the Katherine Gorge road leading to Katherine Airport and paid a visit to the Flight Service Officers at the Katherine Flight Service Unit. I met the OIC, Don Middlemiss there who was working on a day shift and he showed great interest in my latest transfer and provided some excellent advice about the road to Alice Springs and some of the people I should contact. Inside the Flight Service Unit, officers Jim Brady and Don Hodder were busy on the station morning shift, transmitting and receiving messages on the radios to and from aircraft flying all over this part of the state. A military exercise was in progress "up-country" and they were busy coordinating traffic position reports from both military and civil aircraft. It seems through history that not a lot had changed! I said a quick hello to everyone and set off down the track again.

    I arrived at Mataranka, where the railway line from Darwin passes on its way to Larrimah and stopped to have a quick look at the road turn-off to Elsey Station. This was where Mrs Aeneas Gunn wrote her famous autobiography, "We of the Never Never" in 1908. When you have read the book it is interesting to see the site in real life. The Stuart Highway was barely a single lane wide in the early 1970s and driving a small sports car became quite hazardous when encountering long road-train transport vehicles. These huge diesel-powered trucks sped along, towing up to three or four long trailers loaded with cattle or farming produce. They sometimes drove too fast and snaked from side to side with the multiple wheels occasionally running off the bitumen sealed section onto the rock strewn road shoulders. Most of the time I was literally forced off the road. Common sense and basic survival instincts made me pull off into the bush as soon as I saw the tell-tale cloud dusts up ahead that told me a road train was heading my way. Overtaking or passing opposite direction traffic was impossible because the road edges were too rough for the tyres of domestic type vehicles and punctures were guaranteed. When you drive past Larrimah, the scenery changes quite abruptly from tropical undergrowth and heavily wooded country to open scrub and almost featureless arid plains. The deeper you drive into Central Australia, the more arid the environment becomes. Great care had to be taken to maintain a vigilance for possible animal movement across the roadway. Animals caused many fatal accidents both during the day and night and caution was very much the order of the day. Despite the temptation to travel at high speed to try and shorten the trip. Alice Springs is normally a minimum of a 12 to 14 hour drive from Darwin depending on road conditions. Of course, stops along the way are pleasant when refreshments and toilet facilities are available. The hours driving were long travelling in straight lines up and down the undulating sand dunes and bush. As you drive over the next high rise, the road literally disappears over a horizon many kilometres away. Mirages often appeared in the dip ahead, giving the illusion of water running across the road. The shimmering heat often distorted the horizon to such an extent that you began to suffer from quite a headache from the glare. As you approached each mirage, it vanished as the angle of vision changed and then reappeared further ahead in the next dip in the road. It was quite fun really and relieved the monotony of long distance driving in the inland.

You can imagine my intense surprise, as I sped along the track in my little sports car, when a rather large mirage failed to disappear and worse, as I approached it at high speed it appeared to be not a mirage at all! Unable to stop in time, I hit the pan of water with a gigantic splash that would have impressed even a Hollywood Film Stunt Director. I was amazed. How could water be across the road like this on such a clear day in summer? There had been no rain. The hot engine steamed and protested in the water and I quickly selected reverse in order to get back on the dry road. Another shock! When I looked around, I realised that the water was now a long way behind me and the gap between myself and dry land was widening rapidly. I was in the centre of a flash flood. It must have rained up-country releasing a wall of water down onto the creek courses of the plains below. I didn't have any time to reason out my dilemma and all I could see, was the posts sticking up out of the water on either side of the road. Worse still, the water was rising rapidly and leaking through the car doors and the engine was now clearly flooded! It backfired a couple of times, stalled noisily and then refused to start again. I began to feel the onset of panic and jumped out of the car in order to fold the vinyl roof down. Then I sat on the boot with my feet immersed in the water that now covered the drivers seat. I stared miserably around me, as the muddy water swirled past. I really did not want to die like this, but there was no way out. There wasn't even a tree that I could climb up for safety. The water stopped rising after a while so that I was now sitting in it up to my knees with just the windscreen sticking up in front of me. Even in the heat of the day, the water was cold and uncomfortable and I was soon filthy from being immersed in the swirling mud. I thought briefly that my boss owed me a lot for this favour!

After an unhappy hour, my despair began to fade when I heard the unmistakable sound of a diesel engine approaching. In the distance I could see a truck slowly ploughing through the water towards me causing a bow wave on either side of its bonnet. The driver stopped and with typical outback humour leaned out of his cab and asked me what I was fishing for? I explained that I was actually sitting on my sports car submerged beyond my control. He roared laughing and offered me a tow which I gladly accepted and using his towrope, I foraged underwater to secure the towline on the MGB chassis. I knotted the other end over one of the vertical supports on the back tray of the truck and slowly we set off, with me sitting on the boot again, my feet on the drivers seat and leaning forward to steer the car underwater. I must have been a humorous sight to any onlookers as I surfed along behind the truck. After a few miles my fingers were quite numb from the coldness of the water and my arms were beginning to ache. When we reached higher ground, we found thirty other vehicles stranded, waiting for the water in a gully ahead to recede.

As my submerged car became visible, a mighty cheer went up from the delighted onlookers, some even offering to pay money to participate in the new sport that I had obviously just invented! I was not fully able to share their mirth. Some of them were as bedraggled and as wet as I was. There seemed to be an instant camaraderie from everyone caught in the same situation and we stood around on the highway cut-off from civilisation and

Stuart Highway NT on higher ground after recovering the submerged MGB sports car (left) from the flash flood 1971 (Photo Author)

chatting as though we had known each other for years. Mothers provided young children with biscuits and flavoured water to keep them occupied while the men discussed the road conditions, weather and what the road would be like further on towards the Alice. Some tinkered with the engines of their trucks and cars while some simply sat on the side of the road armed with a small stubby bottle of beer while they contemplated what might happen next. Surprisingly, I discovered that many of these folks were other public servants from work areas like the Postmaster Generals Department (PMG), the Bureau of Meteorology and DCA Airport Firemen, all en-route to a new posting in some remote part of the country. Some had completed their remote posting obligations and were on their way back to the city for a posting break. Flight Service Officers seemed to be posted to more remote locations than anyone else and sure had to put up with a lot of inconvenience! But no one really seemed to mind and a significant team spirit seemed to prevail in those days.

Some hours later, the water subsided sufficiently for us to proceed, although the MGB electrical system was being very cantankerous indeed. Such were the experiences of many DCA staff who were posted to remote locations around the country. From Meekatharra in Western Australia to Normanton in North Queensland, they fought many hard conditions, always with the aim of trying to get through to their stations so that the umbrella of radio communication and airfields could be maintained across the land.

After an overnight stop in Tennant Creek, I set off early and detoured briefly to visit the Tennant Creek Flight Service Unit at the local airport where I met Bob Burrows. Typically, I was offered a cup of coffee and a chat about the weather and conditions around the area, which was all very comforting and friendly. A number of local charter pilots were already in the small briefing office planning their day and they too had time for a friendly interlude before I was obliged to say goodbye and hit the road again. All went well until I started to approach the northern gradients that led to the McDonnell Ranges at Alice Springs. As more hills came into view, the weather unleashed heavy downpours of rain and this concerned me greatly considering my recent experiences. It was not long before I discovered lines of traffic stranded again, waiting for raging creeks in the hilly valleys to diminish so that we could pass by safely. I became submerged again in a creek but some on-lookers helped me out again and so I had endured yet another flood on the way to Alice Springs. I was starting to get used to the idea but fortunately experienced nothing like the surfing experience of the day before. The unseasonal rains had of course been most welcome for the locals, but certainly not to travellers like myself, however much the country needed it. The car was now very sick indeed and on arrival in Alice Springs the electrical system had to be replaced and the upholstery completely steam-cleaned.

My first night was at the Oasis Motel owned by Bernie Kilgariff who had been one of the first Connellan Airways employees, who later became a local NT politician. He was amused by my situation and allowed me to strip the MGB on the motel lawn in order to clean everything up. The area looked like a cyclone had hit it in the middle of the cleaning operation. I pulled my suitcase out of the boot and water poured from inside. Books were ruined and my rudimentary clothing was ruined. Later at the airport I met my new boss, the late Frank Hind, who was very understanding and had been in many such predicaments himself and so he gave me a few days off to recover, fix my equipment, car and laundry! Frank was well known as a local actor and had appeared in some of the "Boney" aboriginal detective films that were screened nationally. Frank had a very logical attitude to everything and was totally fair in all his dealings with everyone even though he demanded high standards of his staff.

I discovered that despite the vast distances of the outback, the people that lived there were part of a very close community and my adventures had already been the subject of gossip on outback radio connections including DCA frequencies. A local pilot stopped for a chat on the airport parking apron and even inquired if I was the new bloke that had been using his sports car as a submarine! I was finally able to begin work on the Flight Service roster and worked at the old airport site located on the western side of the new Alice Springs Airport.

Alice Springs Adventures

The Flight Service Unit at Alice Springs was equipped with some very ancient communications equipment that was located in the old tower building next to the small air terminal. This cluster of buildings had recently been used to make the famous movie 'A Town like Alice,' made from the storyline of the famous Neville Shute novel of the same name that was written in 1949.

The Alice Springs Air Terminal as it was in 1970. The building to the right of the aircraft was the old District Airport Inspectors office. The long shadow on the apron shows the position of the Flight Service Unit observation tower. The swimming pool doubled as an emergency replenishment facility for the Airport Fire crew. The NDB mast painted black and white is in the centre right of the picture. The old Alice Springs to Adelaide railway line can be seen at the top right of the picture.
(Photo Author)

The radio equipment used by the Duty Flight Service Officers was located in the observation cab and was the most ancient that I had ever seen. Banks of switches sat in a very complex arrangement in a raised console to the right of the operator. The officers who used to operate the equipment jokingly referred to it as their personal Wurlitzer Music Organ! Old AWA intercom and telephone systems festooned the desk in front of him. A weather information console hung from the ceiling and contained gauges for reading the standard pressure setting (QNH), wind speed and direction, temperature and humidity values. To the right of the operator was a stairwell that required some agility to negotiate because it was so steep and did not have a handrail! By modern day Occupational Health and Safety standards it was a disaster. For this reason, officers were not too keen to run up and down the stairs. However a manual transfer of operational information sheets, flight plan forms, the standard flight information progress strips and updated weather forecast information needed to be transferred from the briefing office downstairs to the radio operator upstairs. A simple but effective solution was found by making a large hole in the floor beside the console through to the ceiling downstairs. A silver chain was fixed to the edge of the console with a bulldog paper clip on the end and lowered so that it touched the briefing officer's desk downstairs. He simply attached the data required to the bulldog clip and in a loud voice yelled "traffic" from the stairwell below to the operator above. I am sure the Marines would have been proud of us! I was intrigued with the local history of the place, particularly since the old airport buildings were about to be replaced by a more modern complex, including a new control tower, all located on the other side of the airfield.

DCA Flight Service Officer John Scougall operating the old toggle switch operated radio console in the FSU Observation Cab in 1971. This equipment was replaced by a modern facility at the new Alice Springs Airport Operations facility in 1972. (Photo Author)

DCA as a Government Department was highly respected and was regarded within the Public Service as a highly prestigious organisation. I was very proud to be working for them. Morale was very high indeed and the DCA area was almost self-contained and run like a military air base. The staff included six Flight Service Radio Operators, three Flight Service Briefing Office operators, three Communications Officers to operate the teleprinter circuits, three administration staff, twelve airport groundsmen, three District Airport Inspectors, three mechanics, three electrical tradesmen, six radio technicians and an Officer in Charge. An Air Traffic Control Unit was about to commence full operations and with it, would be six to eight Air Traffic Controllers added to the list! But for the moment I was interested in how it had all been working up until now. I was saddened when the old building was left derelict after the Flight Service Unit closed down, especially since none of the equipment seems to have been preserved for museum purposes. By comparison, the new Flight Service Unit was equipped with modern radio units including a new radio console, a Fixed Logic Automatic message Switching (FLASH) computer and a modernised briefing office for aircrew flight planning purposes. The Flight Service radio console retained the old RFDS frequencies so that continued assistance could be provided to Flying Doctor operations on a 24-hour basis. The Alice Springs RFDS base callsign was 'VJD' and used three HF frequencies. These were 2020 kHz, 5410 kHz and 6950 kHz to ensure a 24-hour radio coverage with respect to ionospheric radio conditions. The use of these frequencies was particularly advantageous when an aircraft was unreported in the outback. The outstations could be contacted quickly to locate the pilot and prevent a lengthy search and rescue operation. Emergencies in the outback were common. There was always a radio call being made requesting assistance of varying degrees because there had been a vehicle accident on the track or a stockman injured during a mustering operation. Sometimes the emergency was an aircraft having to force-land somewhere or perhaps a lady having a baby and experiencing medical complications. Whatever the scenario, the liaison between the Department of Civil Aviation Flight Service Units and the various services, like the Northern Territory Police, the RFDS, the Darwin based DCA Rescue Coordination Centre and the Salvation Army was magnificent. When someone was in trouble, there was never a shortage of people ready to assist. One evening, I had gone to bed early and was asleep when the telephone rang. The caller was John Jenkins, the duty Flight Service Officer in the Flight Service Unit at the airport. He advised me that there was an aircraft emergency, which involved a missing light aircraft, believed to have crashed near the Petermann Ranges located due west of Alice Springs and he needed someone to assist him. He advised me that the weather was very stormy and that ground searchers from the nearby Docker River settlement were being hampered by rising water in the normally dry creek beds. While he was talking I could hear the busy radio circuits in the background and I was almost dressed by the time he finished the call. I bolted outside in the pouring rain to my car and quickly drove down Todd Street through Alice Springs and headed towards Heavitree Gap. The weather was foul and the rain became heavier as I drove through the murkiness and I stared through the windscreen, wishing it wasn't so far to the airport. Without warning the headlights dimmed and the car began to falter. It began a series of kangaroo hops before stalling completely and stopping in complete darkness. I knew I was near Heavitree Gap, but there was no street lighting because of a town power house electrical failure and I could not see any visual references to find my way in the gloom. Swearing madly, I jumped out of the car in the pouring rain and lifted the bonnet. I became soaking wet in seconds and then discovered my torch wasn't working. While I furiously checked the plugs and electrical connections by feeling my way in the dark, I became acutely aware of a strange roaring and hissing sound. My instinct told me that I was in dire peril from something that I could not identify. Goose bumps raced up my neck and my imagination began to run amok as I listened intently with increasing alarm. Suddenly, I heard the unmistakable sound of water and then I realised that the sound was fast flowing water in Chinaman's Creek not far away from me. This creek was normally dry and connected with the Todd River which was also usually dry. Panic set in and I knew I had better leave the car and head for higher ground. I decided that I had better try and follow the road back in the dark and make my way back into town. The roaring and the hissing was getting louder and with mounting fear, I began to run. I had only run a few paces when I heard a car engine and saw a vehicle coming quickly along the road towards me. I waved my arms madly and thankfully the driver saw me and as it pulled over beside me, I realised that it was a police truck. The policeman wound down his window and bellowed through the din. "What the hell are y'doin out here mate?" I pointed at my car and yelled that I was trying to get to the airport to assist with an emergency. He stared ahead for a few seconds contemplating the situation then shouted, "well the Todd River is flowing already and Chinaman's Creek is rising fast. If I give you a lift to the airport I won't be able to get back again, we'd better have a look at your car." He turned his headlights towards my engine and we discovered that the battery was shorting and that the coil was dead. Northern Territory policemen have to be very resourceful in the bush and he whipped the battery out and banged it up and down on the road before shoving it back in the engine. He grabbed a wire brush from his toolbox and cleaned the battery terminals like a man possessed. Racing back to the police truck he rummaged around for a few seconds and came back with a new coil in his hand! I could hardly believe my luck, it was like running into James Bond! My car started immediately and yelling my thanks, I set off quickly in the direction of the airport. My heart thudded as I saw the peril I had been in. Chinaman's Creek was now visible in the glare of the headlights and had overflowed its banks extensively with the water flowing at an incredible rate in hissing maelstrom of water. The World War II Dam Busters crews couldn't have done better! I paused momentarily at Heavitree Gap to glance over my shoulder to see if the police vehicle had reached safety. Thankfully, he had made it back to the other side and I could just see his taillights disappearing towards the township. In those few seconds I realised that the water was following even faster than I had first thought and was advancing rapidly towards me. I was close to panic as I flattened the accelerator and raced off into the night towards the airport. Six miles out of town, I was still just keeping ahead of the water, which was lapping the road on my left. This meant that the Todd

Alice Springs floods in 1972 at Heavitree Gap. The normal Todd River is on the other side of the trees in the top of the picture and has flooded across the Alice Springs to Adelaide road running parallel with the railway line in the centre of the picture. Chinamans Creek is flowing in over the railway line from the right of the picture. Flooded, it is about 20 times its normal size and usually flows through drains under the railway line. On the night of the emergency, the author was trying to get through to the airport in the direction of the railway line. (Photo Author)

River had overflowed its banks for quite a distance and that was not surprising since the rain was torrential. Rain like this was very unusual in this arid region and was more like the Darwin monsoonal "wet season." The rain poured down without mercy, but I arrived safely at the Flight Service Unit only fifteen minutes later. John was sitting at the radio console and stared at me in amazement. He listened in disbelief as I told him what had happened while I boiled the kettle for some coffee. As he began to outline the nature of the emergency at Docker River, I began to sneeze violently from being wet through and having to stand in the new air-conditioned Flight Service Unit. We both decided that I should have a hot shower over at the fire station and dry my clothes. It was obviously going to be a long night and the chances of getting in or out of the township past all the water in the morning was looking very doubtful. All the bush airstrips throughout the state had closed because of widespread flooding. Even during the early hours of the morning the RFDS frequencies were buzzing with continuous reports coming in from the outstations reporting accidents that had occurred because of the stormy weather. But even the Flying Doctor was grounded because landing aircraft on flooded airstrips and boggy landing strips was extremely dangerous.

The new modern facility at the new Alice Springs Airport Operations facility opened in 1972. The author shown operating the intercom on the Air Ground radio communications console. The console featured VHF, HF & RFDS radio frequencies. In the background is the briefing officers area equipped with a Seimens three-row telegraphy machine automated by a FLASH computer to replace manual message tape handling. (Author)

John and I split the radio workload evenly between us. Reports came in that an Aboriginal tracker party had left Docker River en-route for the Petermann Ranges to try and ascertain if the aircraft crash report was correct and if so the location of the crash site. A mining company had been using a specially designed Cessna monoplane able to land and take off on short rough landing areas in order to support a ground survey team, but the aircraft had disappeared. At this point, the weather began to worsen significantly and many reports started coming in about flooded homesteads and impassable roads. This seemed hardly possible in the Centralian desert environment and the DCA Rescue Coordination Centre in Darwin decided that there was little anyone could do until daybreak. We remained on standby and monitored all the frequencies in case the pilot attempted to call us.

When daybreak arrived, we could hardly believe our eyes. The country between the airport and the McDonnell Ranges was completely under water. The airport was on slightly raised ground and the buildings were therefore protected from the flood, but a torrent of water poured down the storm drains on the runway edges and made the airport completely unserviceable for any aircraft movements. All proposed flights were cancelled. We tried to telephone the township but all the communication lines were completely severed. Our only link was through the RFDS frequencies with the base in town. I was quite amused at having to use the radio to talk to the townsfolk instead of a telephone and then we discovered that Alice Springs was also cut-off in every direction. We were able to use the HF SSB radio link with Darwin, to relay information about the conditions in our area and also to receive updated briefings concerning the status of the search. Clearly we were stranded at the airport for a while and like survivors in a disaster, we sat down and began to ration out the available food items and water. There was not enough for more than a day and the situation looked rather uncomfortable. We were surprised a few hours later when the door burst open, revealing a rather tubby gentleman wearing black shorts and boots, with a bright orange safety helmet perched on top of a mop of bushy ginger hair! He announced that he was from the town rescue service and had come out to rescue us. We looked out of the window at the vast expanse of rapidly flowing water and tried to figure out how he had managed to drive out from the township to the airport. We were unconvinced about the safety aspects of the return trip to town and politely declined his offer. Our rescuer was not to be deterred. He declared that he had not come out all this way to be fobbed off and insisted that we were to return to Alice Springs with him in his rescue vehicle. I reluctantly volunteered to go with them into town to get some urgently needed supplies but John declared typically, that he would remain on duty at the radio console to "keep an eye on things."

We drove out of the airport in a Toyota four-wheel drive utility vehicle and I soon began to regret my decision to go with them. The water looked awesome and I was certainly no long distance swimmer! At the junction of the airport road and main south road we reached the waters edge. I was sitting on the back of the vehicle and was told to hang on tightly. I did not need any second bidding! Before the vehicle had gone more than 200 metres we had become half submerged. I yelled a warning to the driver and suggested that we return to the airport. Still undeterred, he replied that if they could get from town to the airport safely, then they could get back safely. As an afterthought, his companion decided to get out of the vehicle and wade ahead, to see if the road was safe. I regarded this as suicidal since he wasn't using a safety rope or life jacket, but he bravely jumped into the water and vanished. His red safety helmet floated away in the current and he spluttered to the surface in the muddy water and also began to drift away from us. I threw him a rope from the back of the Toyota and helped him back on board while the driver frantically selected reverse gears and shot backwards until we were on firm ground again. I swore violently at them and we returned to the safety of the airport. We spent a quite a few more hours stranded at the airport until the water receded briefly to allow a safe return to the township. On the way there, we found the Alice to Adelaide railway line hanging in mid-air because the embankments had all been washed away. The bitumen highway was also hanging in the air like a black ribbon because the foundations had been completely washed away. I had never seen such a sight! The rescue Toyota that had earlier surprised us with its arrival at the airport had actually driven along a sandbank that had formed on one side of the roadway by the floodwater. However the action of its wheels had actually eroded it away as the vehicle made its way out to the airport. They had been very lucky indeed not to be swept away. Clearly, when they attempted the return trip to town, the embankment was no longer there which was why we began to sink in the water. The rescuers were very subdued when the folly of their actions finally dawned on them. James Bond lived to fight another day! Unfortunately, this was not to be my last experience with inland flash-flooding conditions while I was on temporary transfer duties with Flight Service.

Alice Springs was a curious town in those days, with a transient population caused mainly by employees in government jobs being transferred at frequent intervals. Many others were simply on working holidays or international tourists passing through the area on retirement holidays. One Sunday morning, I was strolling along Todd Street enjoying the crystal clear visibility and deep blue sky. The air was cool and the sun warmed my back. No one could be seen, except for one stranger who ambled into view from around a corner ahead of me. He was dressed in a brightly coloured Hawaiian shirt, chequered trousers and white shoes. A white Panama style straw hat was perched on his head and around his neck hung an impressive array of expensive cameras. As he approached he saw me and called out with an American drawl to ask where he could catch a train into the city centre. Since we were standing in the main street of a very small bush town I was very confused and thought for a few moments before asking him which city he was trying to reach? He became quite indignant and explained that he wanted to catch a train into Alice Springs city. I tactfully explained that he was already standing in the main street, but he thought that I was joking and became quite objectionable so I directed him to the Police station, hoping that someone there could convince him that Alice Springs did not have a metropolitan railway system for tourists!

Alice Springs does not have a lot to offer visitors except tourism and the size of the Australian inland is often difficult for overseas visitors to comprehend. I was amazed one blazing hot afternoon to see two American tourists hail a taxi and ask to be taken to Darwin! The taxi driver explained that Darwin was 925 miles to the north along the Stuart Highway and at least two days travel would be required but they insisted that they wanted to see the countryside! The taxi driver told them that he would not be driving at night but if they were serious he would have to go via his home to get his tooth brush and "tell his missus," and then he would take them to Darwin. I watched the taxi disappear down the street and would have given anything to see how the trip progressed!

I was on the early morning shift the following day and usually arrived at the aerodrome while the sun was rising at approximately 0530. In the winter months the desert areas of the inland are very cold indeed and the formation of ice accretion on car windscreens is not uncommon. The temperatures increased during the day to within 30 degrees Celsius on some occasions. I used to revel in the beauty of the inland winters months because the air was so clean and crisp with visibility in excess of 40 kilometres. When sunrise started the canopy of the sky was black with stars shining quite steadily through the still air. As the first gloriously red fingers of daybreak started, the country side came to life and the spectrum of colours from the bright top of the sun first appearing through to the reddening sky ahead of you merging beautifully with blackness still overhead. It was an amazing sight. If sufficient moisture existed in the atmosphere a fog could occur and the sunrise would be ruined. But Alice Springs airport did not suffer unduly from fogs!

On this particular morning however, I arrived at work to find everyone very excited because Sydney Airport had closed because of fog! I was puzzled as to why that should be a problem for Alice Springs. I then discovered that the reason for the panic was because International aircraft en-route from overseas departure ports such as Singapore, Bangkok, Manila and Jakarta had not been advised of the forecast fog and were past their in flight fuel point of no return. Their fuel situation was so desperate that even Darwin was even out of reach! The only reasonable alternate appeared to be Alice Springs and three aircraft were already diverting there to pick up extra fuel supplies. The Air BP refueller was absolutely ecstatic about this proposition and was busy mustering his staff and equipment to cope with this unexpected invasion.

Unexpected arrivals and Apron congestion at Alice Springs Airport 1972 with a Qantas B707, Alitalia DC8, two RNZAF C130s and a Lufthansa B707 (not in picture) diverted for fuel because of unforecast fog at Sydney (Author)

Alice Springs was not an International destination in the early 1970s and did not therefore employ Customs, Health or Quarantine services on a routine basis for the airport. The Northern Territory Police and Department of Health were expected to perform these duties during such contingencies. A young police officer arrived and informed us that the way he would deal with this onslaught of passengers was not to deal with it at all. In fact, he declared, "No passenger is to disembark any aircraft for any purpose." The airline staff became concerned that the aircraft should not be conducting refuelling exercises with the passengers sitting on board. However, the young officer was undeterred. He declared that he was in charge of the operation and that this was how it was going to be in order to ensure that there were no regulatory breaches. He probably had good reason for being extra cautious but the airlines soon realised that this imposition was quite beneficial for them because they would have little work to do if the passengers were not processed. Within the hour, aircraft began arriving. The first was a Qantas Boeing 707 airliner followed by a German Lufthansa Boeing 707 airliner. A short while later the Italian carrier Alitalia arrived flying a Douglas DC8 with 118 passengers on board. The last two aircraft to arrive were Lockheed C130 Hercules belonging to the Royal New Zealand Air Force. The ground marshallers worked very hard to make enough room for the aircraft to park safely on the tiny Alice Springs aircraft parking apron and the arrivals attracted a great deal of attention from the locals.

There always seemed to be something interesting happening in Alice Springs and the townsfolk of the 70s era were a friendly bunch of people and generally seemed to enjoy a good camaraderie. During my posting I was able to see two rodeos and the Todd River Races. The Todd River was usually dry and teams of people raced bottomless boats against each other by simply picking up the external hull and literally running down the riverbed to a predetermined finishing line in order to received bush prizes of a dubious nature! I remember the local radio station radio 8HA broadcasting a stream of local popular records just before I left in 1971 and one song called "Imagine" by ex-Beatle John Lennon stuck in my mind for a long time for some reason. This was despite numerous other records being played at the time by performers like Chad Morgan and Slim Dusty who were always favourites with the Centralian station hands.

Fred's Flight

My employment as a Flight Service Officer certainly brought me into contact with some memorable people in these outback places. On one occasion, I answered a notice in the local newspaper advertising a tape recorder for sale by attending at the advertised address. A man in his early fifties answered the door and introduced himself as Fred. His rough appearance and belligerent manner concealed a warm humour. He offered me a cup of coffee and I sat chatting to him for a while before I bought the recorder. He told me that he was a cleaning contractor at the airport and knew that I worked there. I soon came to know Fred very well and often socialised at his home. He was a great bloke. On one visit I was amazed to find that a solid brick wall had been built right across the lounge room on the carpet. He told me that his daughter had become involved with a, "long haired, useless good for nothing larrikin," and they had nowhere to live. His wife had apparently insisted that their errant daughter and her boyfriend should move in with them until they could find a place of their own. Fred refused to have anything to do with the boy and decided that he would not live in the same house. However, he compromised and separated the house into two living areas by building the wall and the young couple lived on the other side out of sight. I found his solution to be highly practical although somewhat amusing. On another occasion, he invited me to come around to his house for a beer and watch his newly acquired television set. I protested that Alice Springs didn't have any TV stations, but he snorted with rage in his usual manner and curtly informed me Mount Isa television programmes could be received during freak conditions but a great deal of patience was required! I was very doubtful because Mount Isa was over 500 miles away across the Northern Territory/Queensland border. But we sat for hours chatting, all the time watching random squiggles and lines on the screen on the off chance that a freak condition would eventuate. It never happened while I was there, although we seemed to consume a lot of beer in the process! Alice Springs was scheduled to have its own TV facility late in 1972 and I am sure that Fred would have been glued to the screen all day and probably all night as well. He was a most colourful character and very well known everywhere, especially at the airport, however I had probably not met him because of my shift arrangements and the fact that I was a relative newcomer. While he cleaned the Flight Service Unit, the duty officers delighted in his light hearted banter and slanging matches while they worked and everyone looked forward to his arrival for the day's entertainment while he conducted his cleaning operations. He used to appear wearing a pair of baggy grey shorts, a white T-shirt and a variety of footwear. With a baleful glare, he often arrived in the doorway with a domestic hand-pump cleaning spray hitched by the triggers into the pockets of his shorts like cowboy guns. One day however, the staff thought that they may have been a little too cruel with their banter, because without warning he stopped working and declared that, "if they thought he was just a stupid cleaner and incapable of anything except cleaning then they were sadly mistaken!" He announced that he was going to, "show all you young buggers that I am just as smart as you and can fly too." They watched as he stomped off to the wall, turned the power off to his vacuum cleaner and disappeared out of the door. Appalled at their apparent insensitivity, they watched him march across to the aero-club. He emerged some time later with Ossie, the Chief Flying Instructor and minutes later was seen getting airborne in a Cessna 172 four-seater single engine monoplane. Fred's enthusiasm never waned and began a chain of events that held the constant interest of the Alice Springs flying world over the next few months. Week by week, month by month, Fred continued with his flying lessons and every time he learned something new, he insisted on everyone celebrating the occasion with him at the local pub. He learned to climb, dive, fly straight and level, perform medium and steep turns, stalls and incipient spins. As each sequence was completed, Fred organised a mini-party so that we could all celebrate with him.

When the day of Fred's first solo came, everyone was there to cheer him on. We did not want to distract him and watched from a discreet distance and well out of his sight. We fervently hoped that he would succeed and not come to any harm as the little C172 taxied out. The other instructor at the Aero Club was Rob. He and Fred flew some practice circuits until they both felt that he was ready to go solo. We watched as the aircraft landed after a series of circuits and come to a full stop on the runway. Rob was seen to climb out of the aircraft and walk away onto the grass side-strip. Fred was on his own. After a minute or two, he called the tower and was cleared for take-off. The engine throttled-up and Fred, our hero, charged off down the runway. Safely airborne, he began a turn to the left at the regulation 500 feet and then turned a further 90 degrees on the downwind leg of the circuit so that he was flying as prescribed, parallel with the runway on which he had just taken off. Another aircraft suddenly called approaching from the north and so the control tower called Fred to look for the other aircraft and position himself behind it to maintain safety separation. However, there was no reply from Fred and despite many fruitless calls by the tower, the radio silence continued. Rob was looking annoyed. A radio failure at a time like this was not good for a first solo flight. We fervently hoped that nothing was wrong as we watched Fred zoom around onto the final approach for landing on the main runway in accordance with the clearance given by the control tower.

We heard the Fire Crew ask the control tower if they should "turn out" just in case, but the Controller decided that it was simply a radio failure and everything looked fine. The other aircraft was instructed to widen out his final approach path to stay well behind and land as number two to Fred. Fred executed a textbook landing and as he taxied back onto the Aero Club parking area, a number of people emerged to offer well-earned congratulations. Fred climbed out of the aircraft obviously very happy with himself but we couldn't help noticing that he also appeared unusually tearful. The puzzle remained about the radio. Had it failed? He brightened and a grin crossed his weather-beaten face as he explained that after take off, he had heard the control tower call him to look for the other aircraft. He had turned around to ask Rob if he could see it anywhere but suddenly realised in that moment that Rob was on the ground this time and not in the aircraft. Fred glanced at the empty instructor seat again and then back down at the ground and groaned to himself,

"Hell's bell's, this is the end of old Fred!" He was so overcome, he simply couldn't talk to anyone, especially on the radio. Happily though, he had succeeded and we were all really pleased for him. After some other adventures, mother nature played an unfortunate trick on Fred. Weather conditions in Centralia are fickle, especially in the summer months. There was no mercy given to Fred on the day in question. We were now all used to Fred going out and practicing circuits on his own but on one particular day as he turned onto final approach to land, he encountered some very strong updraughts and down-draughts being caused by a variety of willy-willys and thermal activity. These had not appeared on the weather forecast for the aerodrome. We watched in horror as huge dust clouds violently swirled without warning and engulfed the little Cessna bouncing it around as like a canoe that was shooting some mighty rapids. It goes without saying that modern passenger jets have crashed in such conditions and Fred was about to find out why. As he approached the runway, the aircraft continued to buck and bounce in the turbulence and in the final moments of his flight, he was dumped on the runway like a surf-rider being dumped on the beach. Fred was in trouble. Many of us watched Fred each time that he flew. We all felt like he was our brother and we wanted no harm to come to such a popular local worker. We watched in horror as the elements seized Fred and tried to exorcise him from the air. He valiantly pushed on full power to go around and take off again, but the thermals were stronger than the engine power. Savagely the aircraft was smashed onto the runway with Fred fighting the controls. The Control Tower activated the Crash Siren and it wailed mournfully across the airfield, attracting attention to the fact that an aircraft emergency was now in progress. Staff working in the hangars nearby rushed outside to see if their assistance was needed. The Flight Service Unit began operating on advice received direct from the Control Tower and immediately assessed the flight plans of all aircraft that would be arriving at Alice Springs Airport for the rest of the day. This was in order to determine whether or not operational information in the form of notices to airmen (NOTAMs) was required. If so, this information would enable pilots to assess if they needed to divert to another destination with the required fuel safety margins, in the event of the main runway at Alice Springs being blocked by a disabled aircraft. The procedure required the Flight Service Officer manning the radios to annotate each flight strip of every aircraft that could be affected. While he was doing that task, the Flight Briefing

The end of Fred's Flight, Alice Springs Airport 1972 (Author)

Officer, compiled the text and address list of the operational message that would need to be sent if the worst happened outside. In this way, air safety aspects of all other air operations were maximised so that the chance of an aircraft arriving at the airport while the runway was blocked and with no fuel to go elsewhere was made impossible. Alice Springs Airport has the benefit of three runways, but the smaller runways could have been precluded for use by the operational size of the particular aircraft or even unacceptable crosswind conditions. Mercifully however, the aircraft undercarriage on Fred's aircraft remained intact and after several bounces, he left the runway and bounced across the airfield crashing into a wide storm water drain nearby. The Fire Crew, seeing the drama unfolding, excelled in estimating where he would finish-up and drove at top speed to the assessed impact point. With fire nozzles at the ready in case Fred suddenly burst into flames they watched as he settled in a cloud of dust in the storm drain directly in front of them. Moments later, Fred emerged sweating but unharmed, waving his arms frantically at the fire-crew and yelling "don't shoot!" much to our amusement, imploring them loudly not to cover him with foam. They did not "shoot," but the aircraft was assessed as a "write-off," despite its relatively intact appearance. Fred was escorted back to the Aero Club for a stiff drink and to help write the Air Safety Report. Of course the matter was investigated and put down to the aviation equivalent of an "Act of God." Fred in the meantime remained undaunted and seemed quite keen to continue with his flying training but regretfully, domestic pressures were rumoured and to my knowledge he never actually flew again. None-the-less he won the firm respect of the airport community and typified the courage of outback Australians who are never backward in coming forward to "have a go" at anything, whatever the odds. We all enjoyed many happy times with Fred and I have related this story as a salute to one of the many memorable and colourful people that lived in the Territory at that time. It should be added to the many, that have been recorded over the years in "the Alice."

Dust storm approaches Alice Springs Airport from the Southeast. These storms develop rapidly with little warning and create a major hazard for aviators. The intensity of this storm is much greater than the conditions described in the story above. (PD Parker)

Tiger Moth Tale

Early one morning, I was wandering around the tarmac area at Alice Springs airport, looking at the various aircraft that were parked there. The airport used to have a character all of its own and I am sure the ghosts of men who had flown there are still there. This feeling had a lasting impact on me for a variety of reasons, but today was a day on its own and I reflected on some of the known local history.

The airport was originally part of Undoolya Cattle Station and the land for the airfield was procured by the Department of Defence in 1940. Like the wartime airstrips south of Darwin, the new Alice airfield was referred to by its distance from the main town centre and dubbed the "Seven-Mile." In 1941 it was used by Guinea Airways and the USAAF from 1942. No 57 Operational Base Unit (OBU) RAAF, moved in and administered the facility from 28 May 1942 until 30 April 1946. It was also used by No 24 Inland Fuel Depot, the USAAF 4th Air Depot Group and an Australian Weather Observer/Forecaster Unit. The old Connellans runway was also in use right in the Alice Springs Township and the present day Memorial Avenue is roughly aligned along the axis of the old strip. I recall that late in the 1960's an RAAF Canberra jet bomber accidentally landed there having mistaken it for the new airport located at the 7-mile. This was a somewhat embarrassing problem for the RAAF because the airstrip was not long enough in order to take-off again which required the aircraft to be dismantled for recovery!

The red hue of the McDonnell Ranges stretched east and west of the township and contrasted dramatically with the deep blue sky. The clarity of the visibility was breathtaking and you could see for miles around into the distance. A warm breeze was blowing from the southwest and the day was very pleasant indeed. Nearby an old De-Havilland Tiger Moth (DH82) aircraft was parked in the sunshine. The new white paint on its fabric looked really smart as the old biplane sat as a relic of the past and I pondered what tales that aeroplane would have to tell if it was able to do so.

My dreams were broken by a shout coming from near the fuel pumps on the edge of the tarmac area. I turned around and watched as a lone figure came running towards me waving his arms to attract my attention. He appeared to be in his late twenties and was dressed in an open neck shirt with the sleeves rolled up to the elbows, a pair of light drill trousers and polished brown shoes. He introduced himself as Warwick. He had heard that I was a new aero club member and wanted to know if I would be interested in flying with him in the vintage Tiger Moth biplane that I had just been admiring. I could hardly believe my good fortune because opportunities to fly in vintage aircraft were very rare. I naturally accepted and he told me that he would be flying the aircraft to Bond Springs airstrip the following day. Bond Springs was a dirt airstrip to the north of Alice Springs and normally used by the Alice Springs Gliding Club. I had a rostered day off from Flight Service and we decided to rendezvous at the airstrip at 0830am. I impatiently waited for the next day to arrive and as morning dawned, I set off to find the airstrip and arrived there at eight o'clock. The airstrip was deserted except for the Tiger Moth that was already parked unattended over on one side. The Centralian overnight chill was giving way to warmer conditions as the sun climbed higher in the eastern sky. Warwick materialised from nowhere and enthusiastically showed me over the aircraft. It was actually owned by Captain Christine Davy who was a well-known and experienced inland pilot from Connellan Airways. As the sun became hotter, the scent from the gum trees around the airstrip became stronger. The ancient panorama that is unique to Australia, shimmered all around us, in its own rugged beauty. Warwick sniffed the air and with a grin announced that we should begin flying. Another of his friends arrived and was keen to go flying in the Tiger Moth too, but had to attend another appointment soon and so we decided that he should go on the first flight.

I sat on the bonnet of my car while they prepared for takeoff. The starting procedure commenced with the other pilot standing ready to swing the propeller and start the engine. Warwick leaned over the side of the rear cockpit and shouted out to him, "switches off...petrol on....throttle wide."

He nodded his head at Warwick who seemed to be double checking something in the cockpit, before he leaned out again and shouted

"Throttle nearly closed...all clear...switches on."

The other pilot looked attentive and seemed to know what was going on. Seconds later Warwick shouted, "CONTACT!" and his assistant galvanised himself into action by grabbing the tip of the propeller and flicking it around. The engine started immediately and obviously pleased with his efforts, he ran around behind the wings and clambered up into the front cockpit. He fastened the safety harness, pulled on his flying helmet and adjusted his goggles. They both waved to me as they taxied out and lined the little aircraft up on the airstrip ready for take-off. The engine RPM was run-up to test the magnetos and then the flight controls were checked. With a bootful of opposite rudder to counteract the takeoff yaw caused by the propeller, the aircraft accelerated away, leaving behind a large cloud of red dust which swirled around me for a few seconds.

Author in the front cockpit of the DH82 Tiger Moth at Bond Springs NT 1970 (Author)

I watched enviously as the little biplane climbed high into the crystal clear air ahead until it was at a safe height. It flew around in a wide circle and then rolled lazily on its back into a two-turn spin. They flew around the blue void above for thirty minutes sometimes performing loops and rolls before returning and sideslipping down onto final approach to land. The landing was executed flawlessly and the aircraft floated gently down onto a three-point landing. A three-point landing occurs when the main wheels and the tail touch the ground simultaneously. The crackle and pop of the Gipsy Major engine in the quietness of the early morning was like a beautiful symphony to my ears and I prepared for my turn to fly in this antique aircraft. The motor stopped with a splutter and the two pilots climbed out, chatting happily about the flight.

I climbed into the front cockpit and they gave me a thorough pre-flight safety briefing. The starting procedure was repeated and the engine fired immediately blowing red dust into my face. Warwick shouted at me to give him a thumbs-up if I was OK. I did so and he opened the throttle and we taxied out onto the airstrip. The take off run was only 800 ft or so and we climbed rapidly to an altitude of 1500 ft where I could clearly see several ground features below. To the south of us was the township of Alice Springs, nestling in between Mount Gillen and Anzac Hill. The radar domes at the Pine Gap communications facility were visible to the southwest and I could even see the main runway at Alice Springs airport some 25 to 30 miles away to the south. I was wearing a Gosport flying helmet, which consisted of two copper cups over my ears connected by a length of empty rubber tubing to a mouthpiece just in front of Warwick. I could hear him yelling that he was going to perform a loop.

I gave him a thumbs-up signal to show that I was happy to continue and then gripped the rim of my seat tightly for safety, just in case! We dived to 100 knots and pulled the aircraft nose up rapidly. I stole a quick glance over the left side as Alice Springs disappeared under the aircraft. We zoomed over the top of the loop and closed the throttle. Everything went quiet, except for the wind humming in the wing-wires. We entered a gentle dive and began pulling up into a climb again and resuming straight and level flight. Warwick pitched the nose gently down to build up the speed to 95 knots for an aileron roll. As we entered the climb he pushed the left wing down and the aircraft began to roll. The rudder was used to stop the nose from dropping down as we watched the horizon rotate around us. The wings passed the vertical position and we continued around until we were inverted. At this point the stick must be pushed forward to hold the nose up in order to stop it dropping down suddenly into an unwanted dive. We finally arrived the correct way up and flew straight and level again. The smell of petrol, oil and the dope used to stiffen the aircraft's cloth skin was noticeable as we flew around and enjoyed the view. The Tiger Moth was a real lady of the sky and a pleasure to fly.

We continued with the aerobatics and watched the Centralian landscape tumble around us like a kaleidoscope but all too soon, our fuel supply dictated our return to Bond Springs. Pre-landing checks are completed by making sure that the fuel mixture control is in the fully rich position, the fuel tap is wide open and the friction nut on the throttle is adjusted for landing. With the throttle retarded and no engine noise, I could just hear Warwick calling Flight Service on the area VHF frequency of 122.1mHz and telling them that we had finished our operation and instructing them to cancel search and rescue alerting on us. I hadn't realised it at the time, but for safety he had nominated a time with them, known as a SARTIME (Search and Rescue Time). If we failed to report by that time, then Flight Service would initiate search procedures in case we had suffered an accident. I was really impressed with the fact that someone was actually looking after us while we were flying and concentrated on the view ahead as the airstrip rose to meet us. With a slight skip on touch down, we rolled to a stop near my car. Before stopping the engine, Warwick checked the switches for a rev-drop on the tachometer to ensure that the two magnetos were still functioning properly. Then holding the stick fully back, he cut the switches and when the engine stopped, he pushed the throttle wide open again to purge the fuel lines. The throttle was closed again and the fuel cock placed in the 'OFF' position. We both sat in silence, savouring the last moments of a very enjoyable flight. The engine began making little 'tink-tink' noises as it cooled down after its labours. The moment was over and Warwick climbed out with a grin. The little aircraft bounced up and down on its undercarriage as he jumped off and walked around whistling happily to himself. Reluctantly, I too climbed out and hopped off the wing onto the red dirt again. Nearby two large black crows squawked as they sat in the top of a large gum tree. The Tiger Moth was readied for its homeward trip and with a wave goodbye, Warwick taxied out and took off towards Alice Springs.

    I stood watching enviously as the little bi-plane became gradually smaller until it finally disappeared towards the south. Silence descended around me in the heat of the day and the red dust settled again on the airstrip. I sat wondering why we needed cities and cathedrals when I can sit in the outback and experience this? Suddenly, I felt very alone and started heading back along the Stuart Highway in my car towards Alice Springs. Signposts along the way showed aboriginal place names like Yuendumu, Narwietooma and Papunya and I felt a certain sadness for those who congregate on the coastal fringes near the capital cities and miss out on the pure joy of this outback beauty. Life in the cities is easy and artificial where entertainment and materialisms are all encompassing and seem to promote lust and greed. It seems that a country, which is so fragile, is blatantly ignored and abused. Yet, here in the outback, there is purity in the harshness that can only really be understood and appreciated by those who live here. Here people are far more aware of our basic human survival needs and seem better able to cope with the catastrophes of flooding, earthquake, bushfires, storm and tempest. Unlike the city folk, all races of people who have settled in the outback, are undaunted by the lack of luxuries and take life far more easily in their stride. To me, they are the very heart of this country and the essence of the original pioneering spirit. I learned a lot from them during my charter flying days when I visited cattle stations and other lonely outposts.

Before the arrival of the European explorers, Aboriginal names like Junkata, Gunia, Wadi Jalyur, Anatjari and Mamuru were fairly common on this continent. As the missionaries moved inland, Aboriginal names were partly replaced by biblical names and Joseph, Mary, Ruth, Naomi, David, Abraham and even Goliath began to become common. When the cattle stations began to appear, non-indigenous names began to appear. The cattlemen were a tough breed and called their mate's names like Fred, Trapper, Dingo, Ford and on one occasion I met a bloke called Tractor! I was once told of a family up in the north of the territory whose surname was Motor. There is nothing unusual about that, except that they allegedly had called their two boys Diesel and Petrol and their little girl Lektric although I was never able to substantiate the story! Flying on days like today opened up new vistas of thought as I continued along the highway and I found myself on a spiritual awakening about how the ghosts of the Kyeema had seemingly marshalled to entice me into all these new experiences.

Charlie Mike India

On the morning of 20 January 1972, I was working in the DCA Briefing Office at Alice Springs Airport. This date would have a profound effect on many of the employee's working there including myself. The briefing office usually opened at 6am, however the briefing staff were on duty at 5am in order to compile Pre-Flight Information Bulletins (PFIBs) and collate weather information from the Bureau of Meteorology next door. Pilots were usually waiting outside for the doors to open so that they could come in and make up their flight plans for the days activities.

Photocopying machines did not exist at that time and each pilot was required to insert a piece of carbon-paper between two flight plan sheets so that the "carbon copy" could be submitted to the briefing officer for assessment and acceptance. The Briefing Officer ran a spot-check on times and headings, altitudes to be flown and the fuel calculations in order to ensure that errors were minimised. The flight plan was a fairly complicated sheet to compile depending on the day's activities. Some pilots were lucky and didn't have to write much because they were only going to one destination for the day. However, others with multiple destinations such as the mail runs were obliged to complete quite a few extra details by comparison. Inevitably a queue formed at the Briefing Officer's counter while he worked through each one being submitted. As time permitted, he would place the flight plan on a teleprinter and furiously type out the details that required transmission to other Flight Service Units and Air Traffic Control Units.

For each flight plan handed in, the Briefing Officer also checked to make sure that the pilot had the correct weather details and the latest copy of the Pre-Flight Information Bulletin. Some pilots were quite shoddy in the presentation of their flight plans. Others were highly meticulous and took great care and pride in what they were doing. In Alice Springs at that time, the majority of the pilots coming into the office in the morning were employees of Connellan Airways, SAATAS (South Australian and Territory Aerial Services) or from the RFDS and the Aero Club. One of the pilots who came in that morning was flying the Connellan Airways twin-engine Beechcraft Queen Air to Ayers Rock on a charter service. This trip was relatively simple compared with some of the others, however the pilot was one of the meticulous ones who even apologised to me for smudging the carbon paper slightly on one side of the form. I remember commenting that if all the other pilots took as much care with how the flight plan had been written out, my job would have been very easy. We all knew each other socially and the banter was always friendly. Shortly afterwards I wished him a good trip and he left the office to pick up his passengers and fly them to Ayers Rock. The flight plan normally included an en-route position report on the direct track abeam Tempe Downs station. At 0733am, we heard the aircraft taxi past the office and the pilot calling the tower for his clearances. Ten minutes later, I was standing near the radio console when the duty FSO acknowledged a call from the control tower indicating that Charlie Mike India had departed for Ayers Rock. There was no need to send a departure message out because Ayers Rock was located inside our own area of responsibility. The departure was simply annotated on the flight progress strip with the time we expected him to call at abeam Tempe Downs before going on to report an arrival at Ayers Rock airstrip. However, a few minutes later at the expected radio transfer from the Alice Control Tower to Alice Springs Flight Service the aircraft called but the transmission was accompanied by a very loud audible beeping noise. The beeping was a cockpit warning to the pilot in command indicating that with one or both throttles retarded that the landing gear was not locked down. Given the circumstances facing the pilot who was already dealing with a fire in the right hand engine and a catastrophic structural failure in the wing the activation of the warning horn is not surprising.

The aircraft was no longer transmitting on 122.1 MHz and the FSO tried to re-establish contact without success. As he called the control tower to ask them to try and contact the aircraft on ATC frequencies, we heard the Aerodrome Crash siren activate and a terse comment from the tower on the intercom that the aircraft was returning on fire. We were stunned. No one moved for a few moments then the pilots that were still in the Briefing Office rushed outside onto the airfield to see what was going on and provide any assistance required. We sat down and activated our emergency checklists in concert with the activities already being conducted by the Control Tower. A little later we cautiously asked for an update and advice was received that the aircraft had not made it back to the airfield and had been seen going down smoking somewhere to the southwest. The Airport Manager Alan Miers drove down the road in the DCA Aerodrome Emergency Command Post vehicle accompanied by the Airport Fire Crew but remained in radio contact with the airport. Some time later, he called to advise that he and the District Airport Inspector were on scene at the crash site with the Airport Fire Crew. We prayed for good news but not long afterwards Alan's deep voice came over the airwaves to advise, "…there are no survivors."

I had a hollow feeling in my stomach and I picked up the pilot's flight plan from the pile received that morning and stood staring at it. It was all beyond belief. No-one said anything. All that could have been done had been done. This was my first encounter with a fatal aircraft accident and within hours, Crash Investigators arrived in Alice Springs to begin their inquiries. I was one of the Flight Service staff who was detached from my roster and assigned to assist the Investigators in any way possible. They began by conducting a thorough investigation of the crash site but local Police guarding the area were having difficulty with large numbers of sight-seers who were coming out from town once the news had spread. The situation became so bad that tourists were ignoring pleas from emergency services authorities to stay away and even started trying to tramp around the remains of the aircraft and its occupants. The Chief Investigator solicited the assistance of the Airport Fire Crew who provided a spare crew and tender on permanent duty to sit on the track and chase offenders away by spraying them with the water cannon. Fortunately, these actions had the desired effect and the tourists stayed away allowing the Investigators to get on with their work. The following weeks were sad, tedious but interesting. I had never seen a crash investigation team at work before and could not believe how painstakingly thorough they were. A small field team spent hours at the site walking along the ground on the same axis as the final flight path in order to locate everything that had fallen from the disintegrating aircraft until the final impact point. This included tiny paint flakes found in the sand through to the engines and larger pieces of wreckage. The crash team took over the airport operations room and put up a huge map grid along a complete wall in the room. As each item was identified in the field, it was plotted on a map grid and then brought in to the ops room to be re-plotted on the bigger map. In this way the investigators could see what had failed and fallen from the aircraft first as the aircraft travelled along its final trajectory. The larger pieces of the aircraft were laid out on the hangar floor and intense studies were conducted of all the components. These included metallurgical tests to ascertain likely temperatures of the components, the mode of failure, the maximum temperatures experienced and the duration of heating and the presence of any pre-existing damage. The final report concluded that shortly after take-off, the starboard engine sustained a massive internal failure, which resulted in a large quantity of oil and fuel vapour being released into the engine compartment and then igniting. The fire was able to burn through the side wall of the wheel bay and enter the rear nacelle area where a further supply of fuel was made available by the heat breaching of the oil tank and fuel lines. The intense fire significantly weakened both the main and rear spars holding the wing in place and the complete wing section separated from the aircraft complete with the starboard engine. Incredibly, the time interval between the initial engine failure and the separation of the starboard wing was less than 2 minutes and 21 seconds. The report was released by the DCA Air Safety Investigation Branch in September that year as Special Investigation report 72-2.

There was strong evidence to show that the pilot in command had acted very quickly and efficiently to shut-down the starboard engine and activate the fire extinguishing system. But the tragic circumstances of the chain of events that morning had already overridden any hope that he and the other occupants may have had for survival. It was even more tragic given that three of the

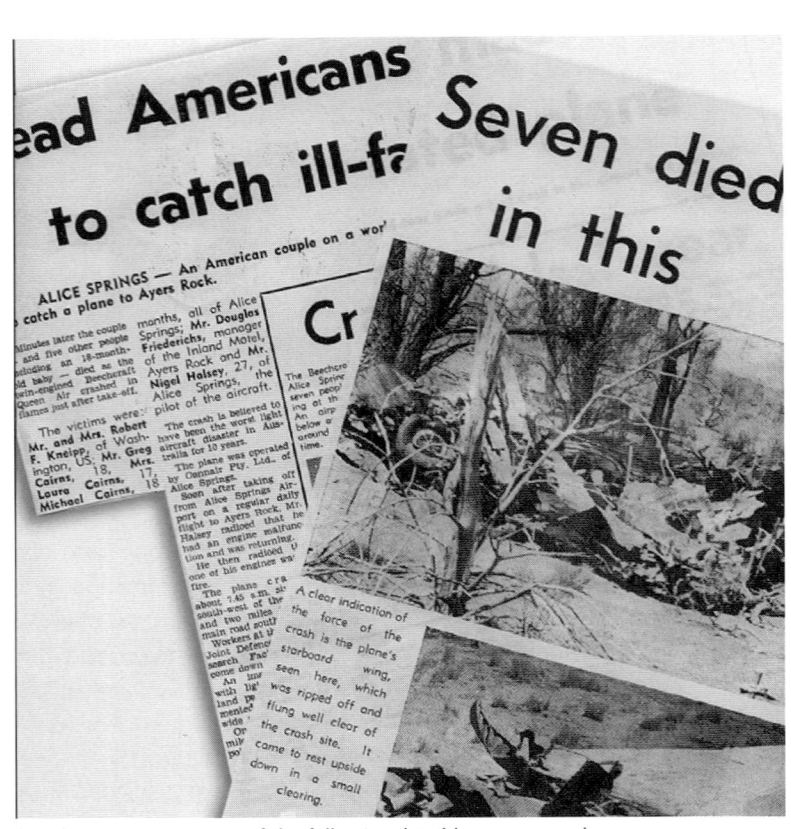

Local newspaper report of the following day. News paper unknown

occupants were directly related to one of the Connellan Airways Engineers and two of the occupants had made a last minute dash to the airport in order to catch this flight to Ayers Rock. A further interesting fact was that Mr E J Connellan had previously notified the Board of Directors for the airline that he considered the Lycoming GS0540 engines to be over-developed for the type of work being conducted at that time. He advised that any advancement towards using such engines was not recommended and on 13 Dec 1969 the Board had recorded this aspect as policy. However, in 1970, there was an urgent need for a feeder aircraft larger than the E50 twin-engine Beechcraft Bonanza aircraft and permission was sought to import Queen Air 70s instead of Queenair 80s from the United States. Mr Connellan was refused permission for such a purchase by the Australian Authorities but curiously was permitted to purchase a Queenair 80 that was the only option available within Australia. His fears about the GS0540 engines used in the Queen Air 80s were tragically realised, although no liability could reasonably be laid against him or his company. (Chap 7, Failure of Triumph by EJ Connellan 1992).

The funeral for the young pilot Nigel Halsey, was held in Alice Springs a few days later. The motorcade was extensive and after a church service held in the township it wound its way along Wills Street turning onto the Stuart Highway. It slowly progressed with NT Police Officers stopping all traffic to allow the sad procession to turn into Larapinta Drive before turning into the cemetery on Memorial Avenue. As the hearse passed by, each Police Officer stood smartly to attention and saluted the fallen airman. The airport must have been running with a tiny skeleton staff if the turn-out for the funeral was anything to go by. Many of us adjourned to the Memorial Club on Todd Street for a stiff drink. Pilots, Ground Staff, Flight Service Officers, Air Traffic Controllers, Engineers and Technicians stood like a lost band of brothers wondering what else Alice Springs Airport had in store for all of us.

Beech 65-80 Queen Air Aircraft VH-CMI at Alice Springs Airport in happier times. (PD Parker)

Oh, you Fly?

Outback humour always seemed uncomplicated and I remembered this story after a visit to an outback airstrip in the Northern Territory while I was waiting with another pilot to pick up a cattleman and take him back to Alice Springs. A local Police Officer drove up in his police four-wheel drive truck to wait for another aircraft that was dropping off some documents for him. While we sat on the fence chatting, a local station pilot entered the circuit area and we idly watched as he made an approach to land. The aircraft seemed to buck and weave all over the place as the pilot struggled in the thermals and mechanical turbulence to stay on final approach for a landing on the airstrip. The wings rolled from side to side and in one minute the aircraft seemed to be climbing rapidly and in another minute it was diving steeply at the ground. The Policeman shook his head in disgust as he watched the approaching aircraft and I too began to wonder if I was going be around for yet another aircraft mishap. The engine noise alternated as the pilot clumsily opened and closed the throttle while trying to fly the correct airspeed for a landing. Thermal activity in the outback is often very strong, waiting unseen to wage mischief on unsuspecting pilots.

We watched with some amusement as the aircraft made a perilous landing and bounced all over the airstrip before finally coming to rest in front of us like a snorting bull in a shower of dust. The engine stopped and the embarrassed pilot climbed out covered in sweat and quickly checked the machine for damage. He wandered slowly towards us, shoving a battered hat on his head and greeted us cheerily. His attire included a pair of well-worn cowboy boots, a pair of dirty jeans and a torn chequered pattern cowboy shirt. There was no doubt that he was a station-hand from one of the local cattle stations. The constable eyed the newcomer up and down and nodded his head towards the parked aircraft. "I liked yer landing mate, I could'na done better meself." The station-hand seemed pleased. "Oh you fly?" he inquired. The policeman jumped off the fence with a deadpan expression oh his face. "Nope" he grunted, and ambled off to get his hat from the truck. The Station Hand scratched his head and ambled over to me. "If 'ee don't fly, 'ow can he do better than me?" I smiled suggesting that he figure it out and I left him there with a puzzled expression on his face watching the Policeman who was now sitting in his truck listening to the Police radio.

Police in the country needed a good sense of humour in their line of duty albeit a dry one. They usually kept a lookout for light aircraft in case of any trouble. If an aircraft was reported missing, but had been seen somewhere en-route, the search could be started from that location instead of from the original departure point. If the outback police stations received notification from aviation authorities about missing aircraft, they always seemed to know where to start looking, or who to contact and often prevented a costly search by quickly locating the offending aircraft. All aircraft operating in remote areas were required to carry an HF radio, capable of communicating with the area Flight Service Unit. Radio communication had to be maintained from taxying at the departure point until the aircraft landed safely at its destination so that any emergency could be reported immediately. The aircraft were required to report regularly along the intended route, so that if it failed to report at a nominated position, the search area would be started from the last position report received. Outback Police Officers could always tell you if an aircraft had been seen in their local area. They always knew what time it had been seen, what direction it had taken and could provide a description of the aircraft type. If an aircraft was not equipped with an HF radio, pilots could still fly within the remote areas, providing they carried an electronic VHF survival beacon (VSB), that activated on impact if the aircraft crashed.

Pilots using this procedure nominated a time by which they would report a safe arrival by using an RFDS relay point or normal telephone. All pilots listened periodically to the emergency frequencies of 121.5 MHz and 243 MHz in case a beacon had been activated. Any pilot hearing an emergency signal was required to make a mandatory "Mayday" call and report the circumstances. The time nominated by the pilot for reporting safe arrival was known as a SARTIME and the Flight Service Unit concerned, annotated the details on the Flight Information Board Flight Strips where it remained under permanent surveillance until the pilot advised cancellation. If numerous SARTIMES had been received by a particular unit, they were arranged in order of time, with the first aircraft expected to report in, being placed at the top and the last one requiring action, at the bottom of the board.

If a pilot failed to report in by the time he had nominated, the Flight Service Unit was required to declare an emergency by an official declaration of a search and rescue phase. This declaration was transmitted by teleprinter to the Rescue Coordination Centre (RCC) responsible for the particular area and they in turn began systematically telephoning locations along the intended route for possible sightings. However, outback telephones were non-existent especially at night time

A bush airstrip in the outback with author taking an accurate bearing on the strip direction, location and characteristics. (Author)

when local exchanges had closed down and if contact could not be made through the RFDS or mining camp radios, then a formal search would begin as soon as possible with aircraft being hired to search specified areas for survivors.

Search pilots always broadcasted as soon as possible, on all the frequencies that may have been used by the missing aircraft in case its pilot was still able to use the radio. If a beacon could be heard transmitting, search pilots immediately performed a prescribed homing procedure to locate the missing aircraft. Most beacons activated on impact so that survivor's dead or alive could be located. Outback Police Officers could always be relied upon to provide accurate information and assistance and always seemed to have a good sense of humour. For this reason whenever the opportunity arose, I constantly visited remote airstrips and recorded a description of their physical details. I usually plotted the geographical position and recorded all the details I could ascertain, including the magnetic direction of the strip, its length, width and surface construction, whether a serviceable windsock was available, the location of the nearest road or track and the closest community that could assist in the event of an emergency. Such assistance included the availability of a telephone or RFDS radio connection. On a couple of occasions when I was concerned about the aircraft operation or sometimes when the weather had deteriorated significantly, my records of bush alternatives where I could land and safely sit out the storm proved excellent on a number of occasions.

On a number of occasions, the Police used Aboriginal men as trackers in the bush. Their keen eyesight and natural hunting instincts frequently proved more than useful when trying to find lost children or injured stockmen. They too had a good sense of humour when you got to know them. I had been on duty in Flight Service one afternoon when one of the local pilots came in and planned a short flight to a nearby station. The trip was only 30 minutes each way to take a Police Officer out in the conduct of his inquiries. He expected to be on the ground for an hour or less before returning to Alice Springs. He invited me to come along for the ride and as DCA had a policy of their officers obtaining whatever air experience they could gain legally whenever they could, I naturally jumped at the opportunity.

Another duty officer was quite prepared to cover the shift for me in my absence and as we all did the same thing for each other, we considered these opportunities to be reasonably routine. All opportunities for Flight Service Officers to become as familiar as possible with the physical features of the Flight Information Area for which they were responsible were to be taken at all times. The pilot had already taken the precaution of ensuring the Policeman was happy to have an extra passenger on board and so I met Jim, who sat in the front of the Cessna with the pilot and I sat in the back looking at the view.

The bush airstrip was actually a long way from the two stations that it appeared to service and we were met there by an Aboriginal stockman who stood grinning happily as we climbed out of the aircraft and ambled over to say the customary "Gidday," His name was also Jim. "What you 'ere for, you lookin' fer something?" he inquired. The Policeman looked down at the ground. "Yeah mate, lookin' for a black feller." The Aboriginal brushed away a fly. His cowboy boots did not seem to fit too well and his jeans were covered in engine grease. His wool blue chequered cowboy shirt was crumpled with a packet of Marlborough tobacco jammed into one of the pockets. "Black feller eh? what's 'e look like?" The Policeman looked up at him. "Not much to go on mate - just black." The Aboriginal burst out laughing. "There's black fellers all over - could be me eh?" I was starting to wonder what was going on when the Policeman also burst out laughing and banged Jimmy on the shoulder with his clipboard. "Jimmy you bastard, you know why I'm here – let's go somewhere and talk eh?" With that the two of them disappeared down a track away from the airstrip and the pilot and I waited in the heat and the flies for forty minutes until their return.

The humour and the banter seemed healthy enough and when they reappeared, the Policeman seemed satisfied with his work and Jimmy seemed no less the worse for wear. The pilot looked at me with a grin. "It's best not to ask anything, they call this men's business in the bush y'know." I nodded in agreement. The less I was told the better it was, as far as I was concerned. We taxied out and called Flight Service on the HF radio to let them know we were on a our way, then if we crashed on take-off and did not call, they would automatically start looking for us ten minutes later. In any case Jimmy was still down at the airstrip and would remain there until he was sure we had taken off safely for Alice Springs. It all felt good and I began to ponder again, the ghosts of the Kyeema and the services that we now enjoyed as a result of their sacrifice.

Search

When aircraft experience a full emergency within the range of an airport radar system, they can be identified by Air Traffic Control and tracked with relative ease. Search aircraft can be directed to the last observed radar position and the survivors rescued with a minimum of delay. However, searches for aircraft operating outside radar coverage depends on the accuracy of the pilots reported position. If the pilot was unable to report the aircraft's position, the integrity of the search relied on other information available. Two types of reporting procedure were used for aircraft flying in Australia up to the early 1990s and were dependent on whether the aircraft was equipped with both VHF and HF radio sets or a VHF set only.

In the 21st century, new ideas have been imposed by adopting airspace models from overseas that demand economic cuts across the Australian Air Traffic Services environment. The models are based on those available from the USA and UK both of which have massive populations that are well able to support such a system. Land use and water restrictions in Australia dictate that we will never have such a population base for at least the next 100 years and such proposals are therefore prone to possible failure. The result of these changes has been the loss of two services that performed to an excellent standard, namely Flight Service and the Rescue Control Centres (RCCs) located in each capital city. These have now been replaced by centralised agencies with the loss of local knowledge that was available in the old units and a lower standard of radio operators who function at non-ATC equipped airports. The costs are imposed on the airline ticket holders and the administrative nightmare facing airports for cost recovery from operators other than airlines is labour intensive to say the least. However, in the halcyon days of the Department of Civil Aviation an aircraft that was equipped with a VHF radio but no HF radio, could still obtain an appropriate safety coverage. The aircraft was able to contact Air Traffic Control or Flight Service Units when it was within VHF range of those units to terminate certain of the Search and Rescue alerting services that were routinely available at that time. Under these circumstances the pilot was allowed to fly from the departure point to a destination without reporting, providing the flight conformed to certain rules. One of the rules was that a SARTIME (Search and Rescue time) was to be nominated and cancelled on arrival, by whatever means was available.

This simply meant that if the pilot did not have an HF radio for continuous communication throughout the flight then he or she could nominate a time by which they thought they would be safely on the ground. They reported by this time in order for it to be cancelled with no further action required by the Flight Service Unit. However, if they failed to report by that time, then the Flight Service Unit was obliged to commence Search and Rescue alerting procedures in order to ensure that the pilot and the aircraft was safe. Pilots who forgot to call in to cancel their SARTIME found themselves the embarrassing subject of an un-necessary search action. The RFDS network or connection by a remote telephone exchange could be used in order to terminate the need for the service for the day. However, in these situations, care was required to ensure that the SARTIME included a buffer to allow for the normal delays in making the contact, so that the alarm was not raised prematurely. If the aircraft failed to report at the designated time and confirmation was received that it had not arrived at its destination, then the entire route between the departure point and the destination needed to be searched. This was a formidable task for the scant search resources normally available because in reality, such an enormous area could not be completely covered within an appropriate time. The situation rarely occurred when the aircraft was equipped with both VHF and HF radio sets. This allowed it to be able to communicate from any location, providing the radios were equipped with the frequencies for the area in which it was flying. If the aircraft was capable of maintaining permanent contact with a Flight Service Unit, then it was normally expected to use a procedure known as full reporting. The pilot was required to make up a flight plan to report at certain en-route positions on a regular basis. In this way, if a position report was missed and the aircraft could not be contacted, then searching procedures were started, commencing from the last reported position report. This significantly reduced the search area because the area from the departure point to the last position received by the ground station did not have to be searched. There is always an exception and the authorities allowed some flexibility for pilots proceeding on a flight of less than fifty miles. If the pilot's geographical location was outside the designated Australian remote areas and continuous radio communications were not possible then a permanent dispensation allowed them to proceed without reporting. This procedure mostly applied to land owners who were taking off and landing within the confines of their property with the intention of carrying out stock and bore inspections. If a search was required for a missing aircraft then the methods of calculating the search area followed internationally recognised search procedures. Once the boundaries of the area had been defined, the search and rescue procedures followed a routine format depending on the type and availability of search aircraft immediately available to the Rescue Coordination Centre.

A typical search was carried out in 1984 for a missing light aircraft in South Australia. The sequence of events started when a Flight Service Unit heard an incomplete and garbled transmission from an un-identified aircraft. The transmission was a terse conversation between the pilot and a second person on board the aircraft and seemed to indicate that the aircraft was in serious trouble. Somehow the radio transmitter had accidentally been activated and the crew were obviously unaware that they could be heard or that they were within VHF radio range of a ground station. The aircraft eventually crash-landed in a remote area and the search that followed was executed with textbook precision resulting in the successful location of the aircraft and its occupants.

The basics of the search are described here to give an insight into the type of activity performed by the Flight Service Unit and the Rescue Coordination Centres of that era. The transmission seemed garbled because the aircraft was making the broadcast just on the edge of the VHF range of the FSU. However the radio transmission sounded like, "...were not going to make it." The aircraft callsign was not used and the FSU had no way of knowing the identity of the aircraft. All aircraft in the area were called to confirm that their operations were normal. Only one aircraft failed to respond and a search and rescue distress phase was immediately declared. The transmission heard by the Flight Service Officer, seemed to indicate that the pilot was on an approach to land somewhere but nothing else could be established. Local commercial radio stations were therefore requested to broadcast requests for the local inhabitants to keep a look out and notify the aviation authorities of any sighting and hearing reports of light aircraft. Sixteen reports were received but intelligence analysis of their reported locations correlated with other aircraft that had already reported satisfactorily.

The following example is a simplified outline of how an intelligence analysis is carried out, using a data sheet in conjunction with a map of the area and a plot of all known aircraft tracks compared with the track of the missing aircraft. If an aircraft was en-route from 'A' to 'B' and was missing, then local radio stations would have been called to broadcast a request to its listeners to advise any sightings of the aircraft to the local Rescue Coordination Centre on the telephone number provided.

When calls were received from members of the public who thought that they may have seen or heard the aircraft, the details were recorded on a table as shown below. The information was then compared with a plot of the intended flight path in order to try and cut down the area that needed to be searched. This task was often quite rewarding because, more searching aircraft could be assigned to a smaller area and this was very gratifying when the aircraft in distress was actually located. The weather conditions did not present any problems for a search, but unfortunately, the incident had commenced late in the day. Daylight was fading rapidly which delayed a visual search from the air being carried out until first daylight the following morning.

The Rescue Co-ordination Centre obtained a copy of the flight plan for the missing aircraft and the search plan was based on the assumption that the only aircraft that was now unreported was in fact the aircraft that had made the transmission which was now causing major concern. Three twin engine aircraft were immediately made available for the initial search actions and were utilised in the following manner.

One aircraft was to perform a night visual search along the track shown in the flight plan of the missing aircraft. The pilot of this aircraft was to monitor emergency frequencies to report on signals received from any survival beacon heard. There was a possibility that a bush fire could be burning if the aircraft had crashed and all sightings of bushfires were to be reported immediately via radio.

The second aircraft was to search in a similar manner to the first search aircraft with the exception that it was to fly two miles displaced either side of the probable track of the missing aircraft.

The third aircraft was to fly a different type of visual search known as a sector search that was based on a ground radio beacon.

The flight plan was analysed again and information based on the performance of the missing aircraft type and the prevailing forecast winds was used to deduce a possible position for the missing aircraft. The missing aircraft was equipped with an Automatic Direction Finder (ADF) which utilises Non Directional radio Beacons (NDB) on the ground to assist in navigation. The search planners decided that there was a chance that the pilot may have tried to use the nearest NDB, particularly since daylight was fading and ground features were becoming difficult to see from the air. The sector search relies on the searching aircraft flying from a datum like overhead the NDB beacon and out to a distance specified by the Rescue Coordination Centre. The required distance from the beacon was calculated to be to a maximum radius of thirty nautical miles. On reaching this point, the search aircraft flew at a tangent to the beacon for five miles before turning back towards it. On reaching the beacon again it overflew and continued on the same heading to a point on the opposite side at thirty miles and repeated the five-mile turn.

Simplified Intelligence Plot

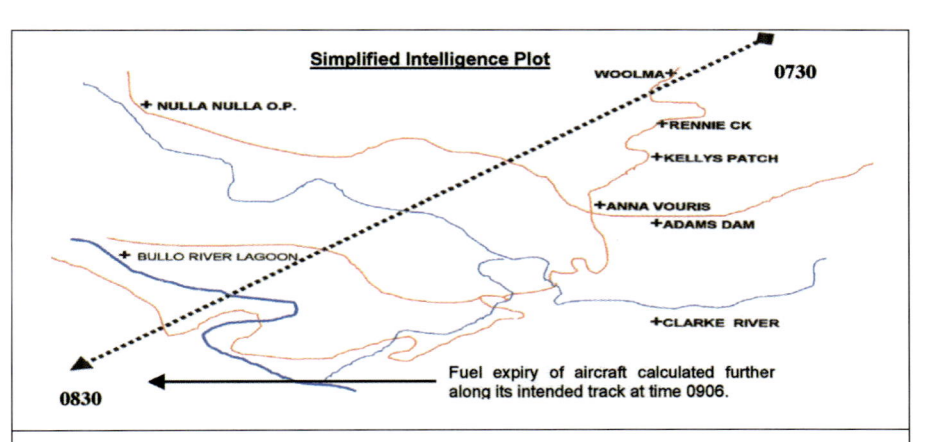

+ WOOLMA 0730
+ NULLA NULLA O.P.
+ RENNIE CK
+ KELLYS PATCH
+ ANNA VOURIS
+ ADAMS DAM
+ BULLO RIVER LAGOON
+ CLARKE RIVER

0830 ← Fuel expiry of aircraft calculated further along its intended track at time 0906.

NAME, & LOCATION	TIME OBSERVED	REMARKS
1. Jan Davies, Woolma Stn	0330-0530	Heard an aircraft pass to the north of Woolma Stn. (Note 1)
2. Ian Smyth, Stockman, Rennie Creek	0600-0700	Heard an aircraft pass NW/SE of Adams Dam (Note 1)
3. 'Doc' Sawley, Nulla Nulla Outpost	0630-0700	Blue white high wing light aircraft pass to the south. Following McKenzie River. (Note 1)
4. Ian Smyth Stockman Rennie Creek	0920	Saw an aircraft orbit the water tanks south of Kellys Patch at very low level & head along McGregors track to the SW (Note 5 and 6)
5. Julie Andrews Clarke Junction Post Office	0810	Saw a red and black aircraft trailing smoke and backfiring travelling in a S direction E of the Junction. (Note 3)
6. Charlie Namatjira Bulloo River Lagoon.	0620	Heard an aircraft go past in the clouds. May have been going to Rennie Station Airstrip. Not sure. (Note 1 and 4)
7. Violet Ngaar Bulloo River Lagoon	0900-1000	Heard a light aircraft - didn't see it. (Note 2)
8. Anna Vouris	1000-1015	Saw a white Cessna 182 flying at about 2000ft heading due West (Note 2 and 5)

- The diagonal black line represents the last known track of the missing aircraft. The aircraft was due to cross Rennie Creek seven minutes after 0730 UTC. Its position was expected to be six miles southwest of Bullo River at 0830 UTC.
- The details recorded in the table can be correlated with the track of the missing aircraft. If the sighting by Violet Ngaar had been more accurate with a description and also closer to 0830 UTC then consideration could have been given to modifying the search area to preclude the section northeast of Nulla Nulla OP to Adams Dam road.
- The report from Charlie Namatjira was discounted because the aircraft appeared to be going in the wrong direction, in addition to being too early for the missing aircraft. The eight hypothetical reports shown here would have presented insufficient evidence to permit the modification of the search area.

129

> **Re: Notes by Intelligence Officer re the table above:**
>
> 1. Occurrence too early to be the target aircraft.
> 2. Occurrence too late to be the target aircraft.
> 3. Aircraft identified as a Beechcraft Baron on a flight from Malvern to Angus Downs.
> 4. Aircraft thought to be VH-TER en-route to Tipperary.
> 5 Aircraft identified as search aircraft VH-CLK
> 6. This report is after the target aircraft's fuel expiry time and therefore discounted.
>
> **Remarks by officer receiving the reports:**
>
> 1. *Anna Vouris requested to ascertain if anyone else had seen the same aircraft & call back with details.*
> 2. *Julie Andrews to be recontacted to verify her report.*
> 3. *The pilot of VH-TER to be contacted to verify the track he flew against flight plan submitted at Cobar.*

By flying backwards and forwards over the beacon in the prescribed manner, the searching aircraft eventually completed a search of an area thirty miles in radius from the NDB. Unfortunately, nothing was sighted on the ground or heard on the radio during this flight. While these operations were being carried out, a high performance aircraft was being prepared for a VHF Survival Beacon (VSB) search. This type of search can be advantageous because it can be performed at a high altitude, which gives a good spread over the ground when compared with a low-level operation.

Since VHF radio signals rely on line of sight, the searching aircraft will be able to hear any Electronic Locator Beacon signal being transmitted in the area contained by the surrounding horizon. A useful formulae can be used for determining VHF signal line of sight range by multiplying 1.2 x square root of the aircraft's height. Thus an aircraft at 6000ft, should be able to hear a VHF radio signal within a radius of 93 nautical miles from its position, provided there are no interference factors such as high ground.

Areas of high ground may present some problems because natural features can shield the VHF transmissions from the emergency beacon. However, flight over the area in straight parallel lines, approximately ten nautical miles apart, will overcome any problems. Emergency beacon signals are distinctive and cannot be confused with anything else.

An aircraft fitted with infra-red heat detection equipment was stationed nearby and requisitioned to assist with the search. An infra-red search area was calculated, utilising the best operating parameters of the aircraft which required long searching runs, because wing bank angles in excess of fifteen degrees automatically shuts off the equipment. Calibration runs were required below the cloud base before the search could start, to seek the most effective temperature parameters. For this particular search an operating range of +5deg C to zero deg C was selected so that items lower than the background temperature could be located. These items included any isolated metal objects like aircraft wreckage. Infra red searches are also normally limited by clouds, areas of smoke or areas of high humidity, but these limitations did not apply during this operation. However, no reasonable target could be located and the next step required the search coordinators to continue planning for the remainder of the night for the visual searches which would be commenced at daybreak. Sixteen aircraft were made available from various air companies in the area and assigned to search areas, which were contained within a promulgated restricted area to prevent interference from aircraft not involved in the search. Throughout the remainder of the night, air-observers, pilots, police, hospital and logistics staff were alerted and commenced work immediately.

Flight Service staff worked under great pressure to provide communications facilities and teletype services for the transmission of flight plans, rescue co-ordination messages, weather reports for the search, intelligence and situation reports. Flight Service staff also assisted in the preparation of equipment to be dropped from aircraft when the survivors were located. Other search and rescue supplies were flown in from the major centres in support of the day's activities. The airport at which the Flight Service Unit was based became a hive of industry, as aircraft began search operations, taking off and landing every two hours in order to relieve exhausted crews, some of whom had become airsick in the heat and the turbulent air. State Police units began deploying in search vehicles and local cattle stations and transport depots were alerted to provide additional support. Local radio stations continued broadcasting search information throughout the period, so that the landowners were kept informed of the events. Two fires had been spotted and were quickly investigated by cattle station staff who were familiar with the area, but they were identified as unrelated scrub fires. During the day a list of available four-wheel drive vehicles and their locations was prepared in readiness for a possible ground rescue. The details were compiled from information provided by the State Forestry Department, State Emergency Service, Police, Mining Companies and various landowners. A rescue helicopter was of limited use, because survivors with internal injuries can be killed by the constant shaking caused by the rotors. One of the search aircraft reported receiving an Electronic Locator Beacon signal and immediately began to home on it in accordance with the recommended homing procedures. Other aircraft also began to report reception of the same signal. In these circumstances great care must be taken to ensure that two separate beacons are not being received simultaneously. One must never assume that another aircraft has not also crashed in the same area. However in this situation, an aircraft was located within fifteen minutes, badly damaged but intact and identified as the missing aircraft. Three minutes later radio contact was established with the downed crew who had managed to recover and operate the aircraft radio. Fortunately, the injuries sustained by the survivors were minimal and the search aircraft quickly identified the most expeditious roads and tracks for ground vehicles to rescue the survivors.

Flight Service commenced relay of the information on behalf of the Rescue Coordination Centre to recall all searching vehicles both on the ground and in the air. Flight Service Officers assisted Search Officers with debriefing of crews and relaying messages of thanks to all concerned. The State Police ground crews reached the survivors using four-wheel drive vehicles and evacuated them to safety and medical care. The cause of the crash was the subject of a separate air safety report. However, the life-threatening situation in which the occupants of the distressed aircraft found themselves, was resolved by the vigilance of the Flight Service Officer monitoring the call and the prompt response of the Rescue Coordination Centre staff available at the time. The aircraft would have been located even faster if the Electronic Locator Beacon had been activated immediately. Another search aspect was the fact that the aircraft still had a serviceable radio that could have been used periodically to make contact with searching aircraft. Finally the aircraft was equipped with both HF and VHF radio sets and could easily have transmitted a distress call that would have immediately alerted the Flight Service Unit to the situation and reduced the time taken, in trying to identify which aircraft in the area was missing.

The Department of Aviation Rescue Coordination Centre at Townsville Airport taken in 1982. This was typical of the standard layouts of these centres. The plotting tables were used by the Intelligence Officer, Search Planner and the Rescue Planning Officer. The Briefing Officers and SAR Log Recorders sat at the side desk equipped with HP85 SAR Computer, Tandberg tape recorders and telephone intercom and radio facilities. The wall data boards were to maintain updated information about search aircraft on and off task together with sighting reports. (Author)

Spencer Gulf

Outstation duties arrived again and this time I was transferred to the Whyalla Flight Service unit in January 1973 in order to release staff for annual leave purposes. I did not realise that my short soujourn there would result in some quite interesting and unfortunate experiences. The Whyalla airport airspace arrangements were simple and consisted of an Aerodrome Flight Information Zone (AFIZ) and this arrangement was generally used at aerodromes serving regular public transport operations and reasonably busy charter services, but not busy enough for the installation of an Air Traffic Control tower facility. An AFIZ was usually designated by a ten nautical mile radius centred on the airfield from ground level up to 3000ft. VHF radio was sufficient equipment because its short-range capability was quite adequate for AFIZ arrangements given the small dimensions of the area of responsibility. HF equipment was therefore not used.

VHF reception was not prone to the static and interference experienced by HF radio equipment, because VHF signals travel in straight lines from the transmitter and do not use the ionosphere. If an aircraft flew over the horizon away from the

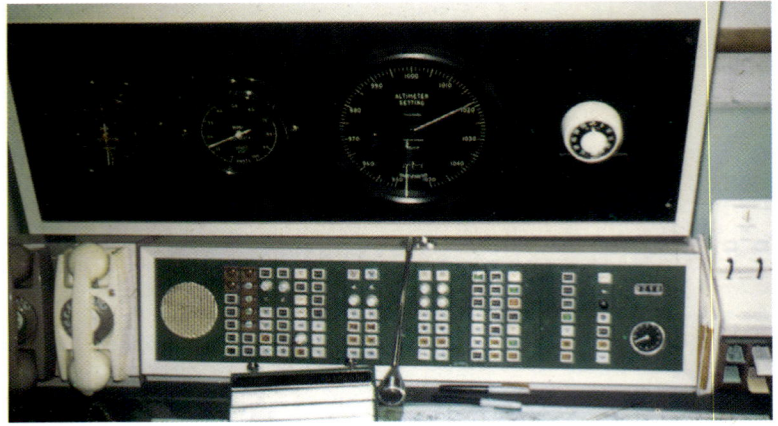

The radio consolette and weather instruments at the Whyalla Flight Service Unit in 1973. The wind is showing 180°M at 15 kts, the QNH setting is 1019mbs. Of interest is egg timer on the right of the top console, used to remind officers about aircraft ETA's to make sure they had arrived safely.

134

VHF transmitter, then it is unable to receive signals from that station. The simple formulae was used by Flight Service Officers in order to find the VHF range of an aircraft by calculating the square root of the altitude in 1000s of feet and then multiplying it by a constant of 1.2, giving an answer in nautical miles. Imagine an aircraft maintaining exactly 1000 feet altitude above ground level. The square root of 1000 is 31.623. Multiply 31.623 by 1.2 and the result is 37.94 which means that the aircraft can usually communicate with a VHF equipped ground station up to a maximum distance of 37.94 nautical miles away. If the aircraft flew 45 miles away from the station then it will be situated just over the natural curvature of the earth and will lose VHF radio contact. Air Traffic became very busy around the Spencer Gulf during the late 60's and early 70's. The requirement for the operation of an Air Traffic Control tower at any of its busy airfields was dependent on their cost effectiveness and the density and complexity of the air traffic problems being experienced. Air Traffic Control presence could often be justified, but the cost of building a control tower as a long-term project was prohibitive. Once a control tower had been built, a decision to close it again was hard to justify. However, the civil aviation authorities constantly reviewed traffic statistics that were maintained for them by Flight Service stations on a routine daily basis.

Australian mining development projects often started suddenly at a particular place, operating for only a relatively short period of time and then closing down again. During their operation however, intensive use of air transportation resulted in amazing air traffic problems at remote locations. At some locations an on-site airstrip was often constructed to reduce the expense of transportation from already established airports. The relatively short time span of some mining operations could not justify sophisticated facilities. This resulted in Flight Service becoming extremely busy because aircraft communications became congested. The town of Whyalla grew as the commercial centre for the iron ore mines operating at nearby Iron Knob. Iron Knob is in the Middlebank Ranges, a range of hills with very rich iron deposits to the northwest of the town. Whyalla itself, is located on the eastern side of the Eyre Peninsula next to the Spencer Gulf and in 1965, the Broken Hill Proprietary Company Limited, opened a steel mill there. The shipyard had been originally built in WWII and grew to become Australia's largest yard until its decline in the late 1970s. Ansett Airlines of South Australia provided Regular Public Transport (RPT) services around the Spencer Gulf using Fokker Friendship FK27 aircraft on daily runs to Port Lincoln, Kingscote (on Kangaroo Island) and Mount Gambier as well as interstate. The tourist developments in the Flinders Ranges north of Adelaide were served by various air charter companies who flew thousands of hours carrying passengers, newspapers and other assorted freight between the major settlements at Woomera, Port Augusta, Ceduna, Whyalla, Port Pirie, Cleve and Cowell. Aerodromes became licensed overnight and the aviation industry rapidly began to prosper.

The usual method of communication for most companies operating in the outback was by HF radio. It was useable over hundreds of miles where telephones had not yet been installed. Its only limitation was the need for the signal to be bounced off the ionosphere and this suffered from frequency fluctuations caused by daily changes in the ionospheric belt around the earth. The fluctuations necessitated the use of high frequencies during the mid-afternoon and the lower frequencies during the night to ensure the signal was reflected at the correct angle back to earth. Radio sets therefore had to be equipped with at least a low and a high frequency to ensure that the reflected signal from the ionosphere always permitted contact during a 24 hour period. Most Flight Service Officers found that a 2MHz or 3Mhz frequency was quite adequate for continuous communications at night and an 8 MHz or 11 MHz frequency worked well during the day. Operators usually used the lowest frequency possible, so that stations responsible for long range routes could bounce their signals further by using higher frequencies without suffering from frequency congestion caused by the numerous aircraft that tried to use the same frequency for radio communications.

Flight Service Officers worked seven to eight hour shifts continuously exposed to the high level of signal noise from HF radio static while trying to listen to aircraft transmissions. Often in later life, they suffered from impaired hearing and premature deafness. Pilots on long-range flights were often reported as missing, because they too were suffering from listening to the same static and had turned the radio volume down for a while. Unfortunately, they were then prevented from hearing a ground station calling for a routine position report and trying to confirm that all was well with the aircraft. When aircraft flew into the radio-range of a ground-station or control tower equipped with the quieter VHF they were relieved to turn the noisy HF radio off! However, VHF signals used in mountainous country proved useless if the aircraft was on the other side of rocky terrain. Mountains in locations like PNG are too high for low performance aircraft and pilots could not contact the ground stations on VHF until they had flown out of the long valleys used for transit and were compelled to use HF to report safe arrival.

As aircraft flew away from Flight Service Units equipped with both VHF and HF radio the normal procedure required a transfer from the VHF short-range frequency to an HF frequency nominated by the controlling unit. Sometimes while en-route, pilots flew past VHF equipped units and although they were not actually flying in the airspace associated with the unit, they were none-the-less in range of its VHF facility. Thus they often called the VHF unit and asked them to relay their position in preference to calling the controlling unit on the noisier long-range HF frequencies. Whyalla Flight Service quite often relayed position reports for aircraft all around the northern end of the Spencer Gulf for this reason.

Three air companies operated from Whyalla in the early 1970s. Desair flew a twin engine Beechcraft Baron (BE55) aircraft (VH-CFO) and a single engine Beechcraft Bonanza (BE36) aircraft. (VH-TYZ) Whyalla Air Taxis used a single engine Piper Comanche (PA24) (VH-PPV) and the Whyalla Aero

Club operated a Piper Cherokee (PA28) (VH-KMO) and Victa Airtourer (V100) (VH-BWI). The St John Ambulance of South Australia frequented the area using a Piper Cherokee Six (PA32) aircraft equipped with a stretcher and the first on-board air-conditioning unit seen in local light aircraft. Trevor Brougham was the Chief Pilot for Desair and also doubled as the Chief Flying Instructor for the Whyalla Aero Club. He was a stocky South Australian in his early forties and balding always with a ready grin. He had a distinctive voice and a very rapid way of speaking. The resident instructor was Mike Carney an athletic looking ex-English policeman and he used to authorise my flights in the club two-seater Victa Airtourer, an aircraft on which I was beginning to build a lot of flying experience. He suggested that I should get endorsements on other aircraft types like the club Piper Cherokee because the Piper was a lot more comfortable on cross-country navigation exercises. I took him at his word and began flying training on the Piper ready for my endorsement by Trevor.

Author at Whyalla 1973 with the aero club Victa Airtourer VH-BWA. (Author)

On the day of the endorsement, my Flight Service compatriots watched as I circled overhead with Trevor sitting beside me and Mike sitting in the back seat. Trevor asked me to depart to the south from the circuit area in order to conduct some more air exercises but at about one mile south of the airfield he suddenly leaned over and deliberately failed the engine. "Ah well," he sighed, "this may happen to you for real one day, so you had better show me that you can land safely without engine power. " I was a bit startled and went through the emergency checklist. On the ground the ever-vigilant Flight Service staff watched as I began to disappear to the south and then realised that the engine had failed. The radio suddenly crackled. "Kilo Mike Oscar this is Whyalla, confirm operations are normal?" Trevor laughed. "Those guys are good - don't miss a trick." He picked up the hand microphone and confirmed that we were on the way back for a practice forced landing. I concentrated madly on keeping the airspeed under control and executed a near perfect landing on the main runway.

Trevor grinned and feigned applause by clapping his hands. "Well old son, on the basis of that, I guess I had better give the endorsement." I was very happy with myself. Mike Carney grinned at me. "See? Told you, you could do it."

Trevor was tragically lost in a fatal air accident at Port Lincoln in the summer of 1974 while flying a routine nightly newspaper and freight run from Whyalla to Adelaide, Port Lincoln, Port Pirie and back to Whyalla again. A flight plan was routinely submitted before each flight showing reporting-points, so that if the aircraft failed to report on the way, search and rescue procedures could be initiated. The Flight Service Unit at Whyalla operated during daylight hours only, because the low traffic density at night did not justify station staffing to cover a twenty four-hour period. However, a VHF link, (one of the first in the country), was installed so that aircraft transmitting on VHF at Whyalla could be heard in Adelaide and vice versa. No such link existed at Port Lincoln. If an aircraft wanted to report landing there, it would have to climb to nearly thirteen thousand feet to communicate over the horizon, using line of sight VHF with Adelaide before landing! This was hardly sensible and so the HF radio was used instead. Unfortunately, the Eyre Peninsular is a poor HF reception area and a popular theory existed locally, that the sandy nature of the terrain resulted in a high absorption of the ground signal, which resulted in a decayed performance. Because of this, pilots were permitted to take off provided they broadcasted their intentions on the ground prior to take off. This was in the forlorn hope that the transmission would be received by someone. In any case, they were not allowed to leave the area near the aerodrome until they had successfully reported a departure to Flight Service in Adelaide. The problem was, that if an emergency occurred before departing, Flight Service could not know about your problem if there was no one else around between Port Lincoln and Adelaide to relay for you. On the night of the fatal Desair accident the pilot had conformed to the usual procedures and reported arrival safely but a departure report was not received within the usual time. No action was officially required by Flight Service, but the officer on duty became suspicious about the delay and telephoned Port Lincoln to see if the aircraft was still there. He was shocked to hear the agent say that the flight had departed a while ago. The Officer commendably, implemented search and rescue procedures immediately and a telephone call was made to Port Pirie, which was the next port of call for the aircraft. The agent there had not seen or heard the aircraft and was still waiting for it to arrive. During this time, continuous calls were made on all possible frequencies including air traffic control. Concern for the safety of the aircraft now escalated, but there was no way a search could be carried out until daybreak. Aircraft operators were alerted to be ready for a search at daybreak. One of the search aircraft that was made ready for a first light departure, taxied onto the runway at Port Lincoln just before daybreak and readied himself for takeoff. He had only been airborne for a few minutes when he spotted aircraft wreckage on a nearby hill. Seemingly, the twin engine Baron had suffered an engine failure and had been unable to out-climb the rising terrain ahead. It had crashed on an upward rising hill surface killing the occupants. Fortunately today, advances in technology have provided microwave links and communications satellites. These permit almost full VHF or UHF coverage all over Australia and HF radios are rapidly becoming obsolete in aviation. However for those on the ground there are still some huge gaps in communications which is why at the time of writing the Royal Flying Doctor Service still uses HF frequencies for communications contact. But during the Aeradio era, Flight Service Officers guarded networks of frequencies across the country on a twenty-four hour basis to ensure no emergency calls were missed. They were dedicated to their work and provided a largely unrecognised though proud service to the flying industry and people of the outback.

I remember Trevor coming into the Flight Service Unit at Whyalla one day a few months before he was killed. He had a charter flight to do a quick trip across the Gulf to Port Pirie and back. The weather was overcast but still amply good to proceed on a Visual Flight Rules flight plan. Trevor departed from Whyalla in the Beechcraft Bonanza and reported a short time later in the circuit area at Port Pirie. While he was gone, I went outside to the meteorological compound on the grassed area just outside the tiny air terminal and wrote out a routine weather report to send on the telex to the Adelaide weather forecasters. Nothing seemed unusual except the sky was overcast. The temperature was mild and the wind was a light and variable breeze. Everything seemed very quiet and there was certainly not much to do in the Flight Service Unit on this particular day. I returned to the radio console and sent the weather report. A few minutes later Trevor called me airborne at Port Pirie coming back to Whyalla. I had already checked with the Adelaide Flight Service Officer to make sure there was no traffic about, since Port Pirie was technically in his area and so I already knew there was nothing about. I advised Trevor that there was no traffic and gave him the weather report for his arrival back at Whyalla. Flight Service Officers do not sit in control towers and rarely had the privilege of a view of the airport on which they were operating. Thus, I was a little startled when Trevor reported arrival back in the Whyalla circuit area and then asked me if I was "going to tell him about the water." I was bemused by this rather odd call and thought for a moment that maybe the weather had started to rain and Trevor was having a dig at me for not telling him. I walked quickly to the window at the end of the office and looked outside at the airfield. I think that my heart stopped dead! Outside, the airfield had become a huge expanse of water. I told Trevor to standby while I rushed outside again to see what was going on. Standing on the apron area I could see that the two unsealed runways had flooded and water appeared to be flowing very rapidly down the western side of the main runway. Trevor was making a very tight circuit around onto final approach for a landing on the still dry eastern side of the sealed runway. As he approached, I noted with alarm that the water was already beginning to cross the runway in some areas and was also coming down the taxiway towards the air terminal. Another flash flood in my life. Trevor landed safely in a spray of water and back-tracked very rapidly leaving the runway via the flooded taxiway back to the hangar. I could scarcely believe my eyes. There was no doubt that I was going to have to close the aerodrome down completely. As I jogged back to the Flight Service Unit, I was joined by the DCA Aerodrome Groundsman, who was looking as startled as I was.

Trevor joined us in the Flight Service Unit and declared that this was something that he had never seen in all his years of flying at Whyalla. I notified the Adelaide Flight Service Centre and they too were bemused by the news. The aerodrome stayed closed for a few days. The water drained safely off the sealed main runway and we were able to use it again for renewed air services. However, the unsealed flight strips were only good for canoeing practice and stayed closed for several weeks! We discovered some time later that when Whyalla Airport was built, the main runway was constructed on an old creek bed. Statistics had apparently showed that the creek was never known to have flowed and a decision was therefore made to use the solid creek bed foundation as a basis for building the runway. Heavy rain further up-country had found its way down to the coast with the result that the creek flowed for the first time known in history and took back its rights to the watercourse on which our runway was sitting! The Eyre Peninsula experienced some severe flooding at that time and even Whyalla suffered flood damage for a short time. I was certainly getting my fair share of floods in my Flight Service career!

Now in the early part of the 21st century, aircraft operators are no longer equipping their fleets with HF radios because they impose a weight penalty on board the aircraft. The least amount of equipment weight is preferable so that a greater payload can be carried with obvious increases in profit margins. The introduction of sophisticated lightweight satellite navigation equipment like GPS has made other traditional forms of navigation obsolete. Satellite communications and cross country microwave links have rendered the Flight Service Units obsolete and the service that used to provide aircrew with a host of personalised services has virtually vanished with little ceremony. Flight Service Officers were used to operating beyond the call of duty, often without payment or recognition. They remained on duty long after their station had officially closed, to ensure that a lone outback pilot arrived home safely. I could never leave anyone out there on his own to fend for himself simply because of union rostering and red-tape. On one occasion, I remained on duty unpaid for three hours, waiting for a pilot who was diverting around some bad storms. I made sure the runway lights were operating and that he had the benefit of VHF radio contact with me. Usually, I advised the responsible capital city centre that I was closing the station in accordance with the rostered requirements and then monitored the radio and weather conditions. I worked with Flight Service staff in many locations and found they were among the most conscientious public servants I had ever seen.

The level of professionalism was imbued in trainees and on one occasion had extreme results. A Fokker turbo-prop airliner, belonging to Ansett Airlines of South Australian was taxying at Kingscote on Kangaroo Island, just south of Adelaide. An over zealous Flight Service trainee spotted a possible traffic-confliction in the form of a De Havilland Drover thought to be heading from the mainland across to the island. The map showed the aircraft were miles apart with no risk of conflict. But the trainee was determined to give traffic information, 'just in case.' I never forgot that very over-helpful voice on the radio!

"Foxtrot November India, this is Adelaide, known possible traffic could possibly be Echo Oscar Oscar a drover who will be departing Penneshaw some time in the future."

Of course the information was totally useless to anyone, and the airline skipper somewhat bemused, replied with:

"Foxtrot November India, thank you...confirm some time in the future?"

The ensuing silence probably indicated that an irate training officer in the Adelaide Flight Service Centre had just pounced on his errant trainee. Nevertheless the incident reflected the enthusiasm which was very much a part of the service. My time came to an end in this region, and I was transferred to Western Australia.

Captain Arthur Turner

I travelled on the Indian Pacific/Trans-Australia railway train from the South Australian township of Port Augusta. This is a two- day journey across the Nullarbor Plain through the most arid looking features of the southern continent. Pioneers laboured here in the most appalling and harsh conditions to build the railway line that now connects eastern Australian states with the city of Perth on the western seaboard. Passing the famous Woomera rocket range to the north, the railway continues westward towards the settlement of Forrest near the WA border. On arrival at Forrest, a large aerodrome with two large bitumen runways can be seen directly behind the railway station. Buildings that used to house the Flight Service Unit and Meteorological outpost in the halcyon years of the 1970s can still be seen as we stroll around while the train stopped here to refuel and take on more water supplies. The local population is very small indeed now and is still dwindling as the requirements for these outposts diminish with the advances of new technologies. Physically, no hills can be seen from horizon to horizon in this vast panorama of stillness and nothing. A few more hours of travel westward along the track we pass the airstrips at Rawlinna and Zanthus as we draw closer to the twin township of Kalgoorlie-Boulder. This twin city is located 595 kilometres east of Perth and here, the Great Eastern highway divides with one turn heading southward towards the township of Norseman standing as the gateway to the Nullarbor Plain and the other turn heading westward again towards Perth.

The Kalgoorlie area was first explored in 1863 by H.M.Lefroy who was looking for reliable grazing land. He led a group of horsemen over 1500 kilometres for three months but knew nothing of the mineral riches over which he was travelling and they went un-noticed until gold was discovered by Paddy Hannan, Tom Flanagan and Daniel Shea at Kalgoorlie in 1893. The frenzied gold rush that followed was amazing and the Golden Mile was born. People endured the most incredible hardships and sickness in the quest for riches. Water became more precious than the much sought after gold and was sold for a shilling per gallon. This would equate today to about fifty dollars! Food was in short supply and outbreaks of dysentery and typhoid saw many people to an early grave, some before they had even reached their mid forties. Walk around the cemeteries in the area and tombstones will be found still standing testimony to the hundreds who died here. The hessian tents soon gave way to battered "humpies" and in time, many of these surrendered to properly organised dwellings. Kalgoorlie boasted 93 hotels in its heyday and even today there are still 25 operating to support a population of nearly 30,000. The Royal Flying Doctor Service still operates from Kalgoorlie airport, although the old airport has now succumbed to time and closed. A new airport has now been built on the other side of the airport road. The Goldfields area is cris-crossed with hundreds of miles of temporary dirt tracks made by various mining ventures. Pilots flying in the area have to take great care not to confuse these tracks with the permanent road systems. When the mine moves on to another area, the road quickly falls into disrepair until the bush claims it again. Survival in this harsh environment dictates that all travellers notify their intentions to family members, their employer, or the local policeman on the track as appropriate. Aircraft were required to report regularly to the nearest Flight Service Unit, or nominate a time by which they expected to complete a flight. A mandatory report by this time was required to avoid the activation of full-scale search and rescue procedures. Microwave telephone links and satellite communication developments from the 1990s onwards, have made reporting in an emergency very easy and so the need for specialised procedures has all but disappeared. Pilots in this era, who forgot to report in to the nearest unit to confirm all was OK, were severely embarrassed while drinking at the local pub, when they suddenly became the focus of the local rescue authorities!

Kalgoorlie is quite green during the winter despite the semi-arid countryside and after the seasonal rains, takes on a whole new oasis-like transformation. Wild flowers burst all over the landscape attracting hundreds of tourists to view the magical splendour. I arrived at night and left the train at the large colonial railway station. I telephoned the Flight Service Unit to let then know I was in town, and also asked about accommodation arrangements. I was pleased to find that I had been booked into a Motel for the first night and could move into rented accommodation situated near the Mt Charlotte gold mine. The house was also only a few hundred metres from the railway line and I could watch the Indian-Pacific train every week as it travelled between Sydney and Perth. I wasted no time in exploring this fascinating place. The inhabitants all seemed to have a refreshingly honest, inlander style of no nonsense attitude, devoid of the materialism's and artificial social attitudes of their kin in the large cities. Most people were directly involved in mining in someway and the airport was one of the busiest I had seen.

MacRobertson Miller Airline, Fokker FK28 type jets serviced the town from Perth every day and on Tuesdays and Thursdays provided two flights from Perth. Added to this, were the commuter services by Trans-West Airlines, Leonora Air Services and charter operations run by Marion Aviation and the Kalgoorlie Boulder Aero Club. The airport was an aircraft enthusiast's paradise. An enormous number of itinerant private business jets and light aircraft from other centres visited Kalgoorlie daily and made the Flight Service Unit a rather busy place in which to work, especially since most of the traffic and operational information services had to be provided on the noisy HF radio.

Morale in the town was high and regular social gatherings at private homes or clubs ensured a high level of interaction that resulted in having to say "Gidday" to just about everyone in the street, because everyone seemed to know everyone else! The memories of that time are most pleasant. The weather is very hot in this region between December and March but the remainder of the year is normally very mild. Rainfall is unreliable with only 225 mm per year falling on average. A fresh water pipeline therefore connects Kalgoorlie with Mundaring Weir near Perth to make up the balance and also supplies water along the 580-

km route, for the many communities along the way like Cunderdin, Merredin and Southern Cross.

Modern aircraft with pressurised cabins and oxygen equipment can fly at high altitudes where atmospheric conditions are much cooler, but aircraft not equipped with such sophistication are confined to operations below 10,000 ft where there is normal oxygen to breath. VHF communication is difficult over large distances because the VHF radio waves only travel in straight lines omni-directionally. When the VHF signal reaches the horizon it continues on into space and cannot be 'bent' to talk to someone hidden by the natural curvature of the earth. Obviously, the higher the aircraft flies above the horizon, the further it can transmit and receive using a VHF radio. The increasing number of jet airliners and turbo-prop aircraft operating at high altitudes were thus able to communicate with Flight Service and other aircraft using VHF over a wide area. This made life a lot easier for the Kalgoorlie Flight Service officers because it reduced their fatigue and exposure to the noisy HF radio. However, all aircraft operating at

The radio console at the Kalgoorlie Flight Service Unit in 1974. The VHF and UHF radios were housed in the green consolette on the left. An AWA CR6B receiver is located at the top left-hand side with the HF radio controls located centre bottom. The emergency alarms are on the right hand side section and weather instruments in the black panel above. (Author)

the lower levels were compelled to use their HF radio sets in order to maintain a continuous radio contact. Radio conditions could be very trying at times while listening through deafening radio static to hear messages from the aircraft. Flight Service personnel were glad when high flying aircraft were nearby and able to relay messages for those flying at the lower levels. We often commented that men had landed on the moon in 1968, but we couldn't communicate with an aircraft only two hundred miles away in 1974! I was on duty in the Kalgoorlie Flight Service Unit one night monitoring the progress of a Flying Doctor aircraft enroute to Esperance located on the south coast. The flight had been requested to pick up a casualty needing urgent hospital treatment at Kalgoorlie. Ionospheric conditions were particularly bad and

An example of an actual Form CA668 as it was then known, used by Kalgoorlie Flight Service in April 1975 to alert Perth Air Traffic Control, and the Flight Service Units at Meekatharra, Adelaide, Port Hedland and Ceduna that an aircraft had not reported arriving at Esperance. The message shows that an IDISWARN is current which indicated to the addressees that a possible reason the pilot did not report was because HF frequencies were subject to ionospheric interference. (Author) the HF radio signals from the aircraft had faded so badly that I could not ascertain if the aircraft had arrived at its destination safely. I tried to telephone the Esperance police station to ask them to drive out to the airport and see if the aircraft had landed safely. Unfortunately, the telephone was not answering. However, by good fortune, a British Airways international jet was flying directly overhead Esperance on its way to Perth from Melbourne and so I called the captain to ask him to relay any messages from the flying doctor pilot. The jet captain apologised, advising that he was in the middle of a call to his London office but that he would return to my frequency shortly. I was quite frustrated. Here was an airliner in my area, talking to his London office on the other side of the world and I couldn't talk to an aircraft only two hundred miles away! When he returned to my frequency, he quickly established that the RFDS aircraft was safely on its way back to Kalgoorlie with the patient on board. If any aircraft failed to report by the prescribed time, search and rescue alerting had to be initiated by the unit first becoming aware of the situation. An urgent search and rescue message, known as a 'Form 668" was completed by the Flight Service Officer on duty with all the details known about the aircraft recorded. The information was then sent via Siemens teleprinter to the nearest Operational Control Centre who responded by activating the Rescue Coordination Centre. This message was also addressed to units responsible for adjoining Flight Information Areas, so that they could monitor the frequencies in order to relay any messages required. The HF radio 'skip distance" varied at night and I often received calls from other Flight Service Units relaying a position for an aircraft operating in my area. Similarly, I could receive a call from an aircraft in the north of the state whose radio signal was skipping over the station he was trying to call and skipping right into the reception of my receivers. I therefore provided the required relay to the station he was calling. Various types of aircraft emergencies were classified into a long list of ground station response procedures. All emergencies were covered, from radio failures in the outback to major crash situations. There were

150

three levels of search and rescue alerting. Each signified a particular situation, so that anyone involved could apply a set of prescribed standard procedures in response to a given emergency. This system also alleviated over reacting to a problem and wasting financial resources unnecessarily. The nearest RCC was notified of a search phase, as follows.

In a time when communications did not benefit from the use of

Stage	Type of Phase & Teleprinter code	Conditions for each Phase Impementation
1	Uncertainty Phase (INCERFA)	Used when doubt existed about the safety of an aircraft. This included failure to report, departing from an airfield, by a nominated position report, when called by a ground station, arrival at a nominated destination, or completion of a flight by a nominated time (known as a SARTIME)
2	Alert Phase (ALERFA)	Used when apprehension existed as to the safety of an aircraft or its occupants. This included loss of radio contact on final approach to land at an airfield controlled by Air Traffic Control, or following an uncertainty phase when no further news has been heard of the subject aircraft, or the aircraft is unable to maintain altitude because of power losses or may experience an abnormal approach and landing for any reason.
3	Distress Phase (DETRESFA)	Used when there was reasonable certainty that an aircraft was threatened by grave and imminent danger and required immediate assistance. This included hijacking, to upgrade an alert phase, or when aircraft fuel was insufficient to continue safely or when the aircraft is about to crash/ditch in the sea.

satellites and microwave repeater links, this system of SAR alerting was economical and worked very well indeed with very few errors. During the entire proceedings, all network stations assisted each other with HF communications and monitored their allotted frequencies to ensure a high level of assistance to the pilot in distress. It is an interesting fact that Flight Service stations were developed from the Australian coastal radio stations that were operating in the 1930s. During February 1934, a de Havilland Dragon aircraft was fitted with a radio capable of communicating voice transmissions on 365 kHz and 500 kHz. The 500 kHz frequency was a marine radio carrier. The aircraft departed from Launceston in Tasmania and headed for Essendon Airport in Melbourne Victoria. As it passed over Flinders Island located in the Bass Strait, the pilot's speech was received for the first time by the Melbourne Coastal Radio Station. Information relating to passengers and arrival times were received and relayed by telephone landline to the airline company office. Before all this happened and developed into a full time air-ground-air communications service, aircrew had been reliant on signals displayed on

Sample of an aircraft movement report message. This was sent was received using Morse Code by the Aeradio Officer at the Forrest Aerodrome on the 10 March 1942. The message is from Kalgoorlie Aeradio advising Forrest that aircraft registration VH-USY had arrived safely and landed at Kalgoorlie at 1603 hrs.
(Civil Aviation Historical Society)

the ground and dropping messages in bottles or canisters. The children's storybook mariners, who became stranded on desert islands, clearly did not have a monopoly on the use of messages in bottles! Sometimes amateur radio operators could be contacted by aircrew using the few frequencies available in the early days and asked to relay a message. Aircraft operators did not take long to realise the full potential of equipping their aircraft with radio installations. With this equipment on board, the aircraft literally became "ships of the air," while they utilised the marine frequencies. This was very convenient to trans-oceanic fliers who could now communicate with ships at sea and so enjoy a certain feeling of protection if an emergency occurred. Of course ships were using known reporting schedules and clocks had to operate reasonably accurately to avoid missing a contact. The Titanic sinking in April 1912 demonstrated this fact quite well.

Arrangements were then made with Australian coastal radio stations for all aircraft to utilise other frequencies using equipment like the Marconi AD36 airborne facility. Traffic schedules were organised and aircraft began to make reports so that the progress of the flight could be monitored. If an aircraft failed to report, then the alarm could be raised and search procedures commenced. Quickly the service developed so that pilots could receive updated weather reports and transmit and receive other pertinent operational messages. In remote locations like the Nullarbor Plain, radio operators were required to use Morse code to confirm the safe arrival and departure of the flight. The Department of Civil Aviation was formed in 1938 and assumed administrative control of communications services, which had been performed originally by Amalgamated Wireless Australia (AWA).

I arrived on a posting from Adelaide to Kalgoorlie during March 1974. The OIC of the Unit at that time was Mr Ken Davies from Perth. The station staff consisted of seven Flight Service Officers, one administrative assistant, a Senior Radio Technical Officer, two electrical engineers and a groundsman. The Bureau of Meteorology was located on one end of the Flight Service Unit building and a staff of six including the Officer in Charge rotated around a 24-hour roster. The staffing arrangements at Kalgoorlie were fairly standard for the size of the township. I was greeted on my first day by the duty Flight Service Officer Mr Clive Cassin who was accompanied on the morning shift by the OIC who was performing briefing duties in the briefing office for another officer who was absent on sick leave. The familiar smell of hot electronic equipment permeated the air and Clive was not using a headset because traffic conditions were not busy. Officers were required to wear headsets when the circuits were busy, to reduce noise interference in the unit and also to ensure that messages were transmitted and received without error and preferably without repeats. The fact that the circuits were not busy was beneficial to me because it allowed Clive an immediate opportunity to show me the various controls on the radio console so that I could acquaint myself with the procedures and unit operation. I waited while he acknowledged a call from Meekatharra Flight Service. They were relaying a position on an aircraft en-route between Laverton and Warburton Mission. He pressed the transmit switch and advised Meekatharra that there was no known traffic conflict in that area. The transmission was business like and brief, "Meekatharra, this is Kalgoorlie, Kilo Alpha Delta, no known traffic." I recognised the callsign 'VH-KAD' which was one of the Royal Flying Doctor Service Piper Navajo aircraft. A brief look at the Flight Information Board, now updated by Clive on receipt of this information, confirmed the aircraft's track. Traffic began to increase again as aircraft began operating between mining airstrips and other outstations. Clive meticulously went about his duties providing accurate traffic information and search and rescue alerting where required. I was invited to dinner at his home that night and he and his wife Ann, told me all about Kalgoorlie and the West. His knowledge of aviation matters was impressive and aviation magazines and books were obviously not in short supply. Ann, like most wives who were married to fanatical aviation husbands, did not share his enthusiasm for flying but demonstrated a practical tolerance just the same. Clive had recently obtained his flying licence and loved to fly the club's Cessna 150 two-seater trainer. The call-sign of this aircraft was Kilo Papa Kilo (VH-KPK) and it wasn't long before the locals had dubbed the aircraft Kilo Papa Cassin instead, because he flew it so often. Strict adherence to radio procedures denied us an opportunity to use the corrupted call-sign operationally, but the nickname became popular and thus he became locally known as Kilo Papa Cassin rather than Clive to the amusement of everyone at the airport. Radio call-signs could present problems occasionally. One aircraft call-sign was Tango Echo Echo (VH-TEE). Transmitted quickly this call-sign inevitably had tongue tied trainees talking about Geckos (small lizards). Aircraft call-signs that were similar sometimes caused confusion. MMA airlines flew a Perth-Kalgoorlie-Perth service and often used an FK28 jet using the call-sign Foxtrot Kilo Delta (VH-FKD). Leonora Air Services based in Kalgoorlie operated a Beechcraft Baron BE55 aircraft using the call-sign Foxtrot Delta Kilo (VH-FDK). The RFDS in Kalgoorlie operated two Piper Navajo aircraft, one of which used the call-sign Kilo Foxtrot Delta (VH-KFD). Invariably these aircraft flew in the local area at the same time and great care had to be taken by Flight Service Officers and pilots to prevent any confusion.

Department of Civil Aviation Flight Service Unit Kalgoorlie Airport 1974. The door at the left is the public briefing office access. Entry to the Flight Service Office is further down the verandah to the right. Note the chimney for the fireplace and radio masts. The flagpole was for the DCA Ensign flag. (Photo Author)

Shift-working arrangements were relatively simple because Kalgoorlie was not required to operate a 24-hour station except in an emergency. Very little traffic operated at night in the area and so responsibility was transferred to the Perth Flight Service Centre until Kalgoorlie reopened in the morning. The morning shift usually commenced at 0500am and the afternoon shift usually finished at approximately 2130 (0930pm). Duties normally started by inspecting the runways for any damage or obstructions. When the inspection finished the FSO then opened the airport briefing office. Most mornings, up to twelve pilots could be found waiting on the verandah outside the office to submit their flight plans for the day. The morning-shift officer contacted Perth to resume responsibility for the Kalgoorlie Flight Information Area (FIA) and any aircraft operating at that time were transferred to the Kalgoorlie station frequencies. The first hour was very busy operating the radios, accepting and checking flight plans and transmitting them on the AFTN teleprinter. Weather reports had to be assessed and updated operational information bulletins had to be compiled in time for the pilots before they departed on the day's flights. Considerable liaison was usually required with other units about aircraft movements crossing the unit boundaries. Pilots from outlying communities also telephoned in with flight plans while all this was going on and obtained briefings about the weather and any operational information pertinent to the flight. Each call normally took about 10 to 15 minutes to complete. Flight plans from other ATC and FS Units began arriving at the same time on the receive teleprinter and cardboard flight information strips had to be written out and placed on the flight information board to ensure that all flights were correctly accounted for. Union rulings dictated that the next officer on duty commenced at 0600 and this officer usually arrived to find the early shift officer working hard on a multitude of seemingly endless tasks. The oncoming morning shift officer was therefore required to provide all possible assistance to his beleaguered comrade with one officer concentrating on the briefing-shift duties while the other operated the radios and

156

assessed flight plans. Finally another officer would arrive on duty at 0800 and the VHF radio was switched from the main console and operated separately in order to reduce the workload and frequency congestion. After the early morning rush hours, the briefing officer settled into an administrative routine keeping records of airfield and unservicable radio navigation aids in the region and compiling Notices to Airmen (NOTAMS). These were typed out and displayed on Pre-Flight Information Bulletins (PFIBs) which listed routes flown set out in geographical order or alternatively set out on a similar bulletin which was compiled in alphabetical order. Each bulletin usually consisted of three to four pages of either foolscap or A4 size paper containing the information and this was usually classified under three main headings namely aerodromes, navigation aids and general information.

In the early 1970s, Flight Service Units were all equipped with an eight megahertz frequency for High Frequency (HF) air communications using 8938MHz. Unfortunately, this frequency soon became congested because of the large numbers of aircraft and ground stations using it. Further congestion occurred because aircraft could not always get through to the unit they were calling. This was often caused by weak aircraft transmitters or simply the radio conditions for the time of day. This meant that units across the country were obliged to spend a lot of time relaying intercepted radio messages backwards and forwards on behalf of other units. Eventually, the problem was solved when different 8MHz frequencies were assigned to different units. HF radio conditions were sometimes such, that an aircraft operating near an outback unit like Kalgoorlie could not be heard by that station but could be heard quite clearly by the Derby FSU because of the 'skip distance' caused by fluctuations in the ionosphere. These fluctuations causes the signal to be reflected back to earth at an incorrect angle thereby completely missing the called station. Mandatory relay procedures ensured that all aircraft calls were accounted for. If radio contact could not be made to another unit to relay an aircraft position report, then the report was sent via the

157

teleprinter circuits. The relay procedures adopted by the ground stations were vital to pilots who were experiencing in-flight emergencies or weather problems.

One in-flight emergency occurred when a flying doctor aircraft was returning to Kalgoorlie with a critically ill patient on board. The RFDS pilots are vastly experienced and are hard to replace if one of them is grounded because of sickness. The RFDS therefore sometimes employed reserve pilots, selected from local flying organisations to ensure the maximum service possible was given to those in need of immediate medical care. The reserve pilot on this occasion was Arthur Turner who had volunteered to do the trip to help the Doctors with their problem. Arthur was the Chief Flying Instructor at the Kalgoorlie-Boulder Aero Club and he knew the local conditions well. The trip was not one that would be pleasant. The weather forecasts showed the most appalling weather covering the state. None-the-less a decision was taken to try and rescue the person who needed immediate medical aid. It was to be a classical flying doctor story. While returning to Kalgoorlie, the aircraft was suddenly hit by lightning that destroyed some of the on-board electrical equipment. As a result of this freak strike, one of the engines began to surge causing Arthur some flying control problems. The descent into Kalgoorlie was by all accounts quite frightening and Arthur began to feel grave concerns for the safety of the flight. He transmitted an emergency call on the radio and fortunately, the VHF radio was still working, but worse was to come for the embattled occupants of the aircraft. The local emergency services arrived at Kalgoorlie airport to wait for the aircraft but as Arthur guided it around the landing pattern, another lightning strike hit the airport power supply causing all the aerodrome lighting to fail completely! Desperately, the nurse on board grabbed a torch and trained it on the aircraft instrument panel so that Arthur could at least monitor the speed and altitude of the aircraft. The power surges in the engines were now so bad that they could no longer climb the aircraft up to a safer altitude and they faced the reality of having to land in the dark by guesswork. The undercarriage system had lost electrical power and Arthur frantically lowered the undercarriage using the manual system. This was exhausting work as he aimed the aircraft at the point where he last saw the outline of the landing area. If all else failed, at least the emergency services were there. In the final seconds of the approach to land, sheet lightning momentarily lit up the runway and enabled Arthur to line up accurately for the landing. His excellent flying skills contributed to the success of the flight while Flight Service coordinated the emergency services and kept the Perth RCC updated.

I became interested in the Kalgoorlie Boulder Aero Club (KBAC) and formally met Arthur for the first time. He was widely experienced in aviation and seemed to have flown everything as a young Flight Lieutenant in the Indian Air Force from Typhoon fighters to De Havilland Tiger Moths trainers and heavy transport aircraft. He was a perfect gentleman who was always ready to help struggling pilot trainees. His instructing was thorough and accurate and he was respected by everyone who came into contact with him. I later joined the club committee and frequently enjoyed a social evening with Arthur and club members to talk about flying. Tragically, Arthur was killed in the 1980s while yet again trying to evacuate a medical emergency from an outback mission to Kalgoorlie for the RFDS. On what was to be his last trip, he had successfully picked up the patient and was on his way back to Kalgoorlie. The aircraft ran into an unexpected and very violent inland storm, only one mile from the main runway. Although the duty Flight Service Officer had broadcast the changing wind conditions, the night concealed the tremendous forces that were blowing in the area. Arthur suddenly found himself sinking in a violent down-draught and could not arrest the high rate of descent before the aircraft collided with a poppet head used by one of the mines on the Golden Mile.

I shall always remember Arthur with respect and have many happy memories of flying with him in the West Australian skies. I completed all the navigation exercises for the issue of my unrestricted Pilot Licence with him and he instilled in me a confidence that I had not known before. His standard of airmanship was excellent in every respect and I could not have hoped for a better or more patient instructor. Not long after my arrival the Aero Club sold the Piper Warrior PA28 and replaced it with a Cessna 172 type of aircraft. My log book shows my first flight in this aircraft was 10 August 1974 during which I flew a local flight around the Kalgoorlie area.

Ten days later I was airborne again for a local flight around the training area in the most beautiful flying conditions possible. It was hard to resist the urge to go flying. I called Flight Service and asked to be put on scheduled times for search and rescue purposes. If I failed to report, they would start looking for me. The sky was a deep blue and the temperature was in the mid-twenties with no haze or smoke anywhere. I climbed to 6000 feet and began to practice steep turns, stalls and incipient spins. The aircraft handled really nicely and I revelled in the calm conditions and the tremendous view around me. The conditions made me remember the day I took my first flying lesson down the Queensland Coast a few years before. After I had been flying around for approximately forty minutes I decided to return to the airfield to practice some circuit flying including touch and go manoeuvres to practice landings and take-offs.

I called Flight Service and reported inbound. They advised that there was no known traffic and provided the usual weather update. I trimmed the aircraft for a standard descent and made sure the fuel cock was selected on the 'both' position so that fuel was given from the two tanks simultaneously to the engine. This was required for safety reasons, so that during the landing, a failure of one fuel line would not result in an engine failure. I pushed the mixture control fully into the panel to make sure the fuel to air ratio was fully rich for the descent and then glanced at the fuel gauges to confirm the quantity of fuel remaining.

My heart nearly stopped! Both gauges were reading totally 'empty.' I was too far from Kalgoorlie airport to make a glide approach to the runways and broke out in a cold sweat. How could this have happened? I began feverishly searching the

160

ground below me for a suitable forced landing area, ready for when the engine stopped, but there was nothing except low bush scrub and trees. The onset of panic began. How could I have been so stupid? Maybe, I didn't put the fuel caps on properly after I had refuelled the aircraft? I was puzzled. I had filled the tanks to capacity before I had left and I remembered double-checking them as I always did before I had taxied out for take-off.

I looked at the gauges again and tapped them hopefully, but to no avail. There was no other explanation, the caps must have worked loose somehow and the fuel had all been sucked out by the slipstream. The problem in this aircraft, was that the high wing configuration prevented me from seeing the fuel caps, because they were both on top of the wings above me.

I picked up the microphone to transmit a 'Mayday' call. I imagined Flight Service were not going to be very pleased with this little effort and my reputation would be tarnished for a long

Kalgoorlie Airport 1975. Taken from 3000ft looking towards the South. Clearly visible at the centre of the picture is the main runway 10/28 used for jets with the shorter light aircraft runway 18/36 to the right. A red dirt airstrip 14/32 was available in dry weather conditions only. The Flight Service Unit and air terminal can be seen at the end of the taxiway leading from the centre of runway 10/28. (This airport has sadly been decommissioned for housing developments, with a new airport constructed further south in the early 1990s). (Author)

time after this! Its not easy making a Mayday call for the first time and I faltered momentarily. What was the correct procedure?

The radio phraseology formed in my floundering brain and I pressed the transmit button but something else was wrong. The radio was dead. This was not my day. First I had run out of fuel away from a decent landing-ground and now the radio was dead. I thought briefly that I might soon be the same way. The situation was not encouraging. Miserably, I transmitted anyway, hoping that even if the receiver was dead that maybe the transmitter was operating properly. I could see Kalgoorlie Airport now and mile by mile I came inexorably nearer. My heart was just about in my mouth which had gone very dry indeed. I kept glancing at the fuel gauges cursing my stupidity and expecting an engine failure any minute now, but the engine purred on as though it was being supplied from a non-mortal source. I kept a good lookout for other aircraft and positioned myself for the nearest runway circuit position. My despair turned to elation as I realised that I was now within safe gliding distance of the runway threshold. I was careful to remain a little higher than normal to ensure that I did not undershoot the approach. It would be a shame to crash only metres from the runway threshold after being given this second chance to escape a ghastly accident. I landed without further incident and taxied up to the Flight Service Unit to let them know that I was OK and also report the radio failure. They would probably want me to fill in an air safety incident report, since I had not been able to call at the mandatory Flight Information Zone positions anywhere. Anyway, I was interested to know if they had heard any of my transmissions made "in the blind."

As I climbed out of the aircraft, I felt chilled by the wind and looked down at my shirt. It was soaked in sweat! I turned around and walked straight into Arthur who had realised that there was something wrong and walked over to the Flight Service Unit to find out what was going on. He listened to my tale of woe and climbed up on top of the wings to check the fuel. I was astonished when he announced that the tanks were still full! He climbed down and began to fiddle with the switches in the cockpit. He

162

climbed out of the door and gave me a rather withering look. "Your stupid clot," he admonished and pointed at the aircraft. "You had an alternator failure and flattened the battery which is why the radio wouldn't work. The fuel gauges are electric, so it wouldn't matter how full the tanks were, they would still read empty if the electrical power to the aircraft systems have failed."

The late Captain Arthur Turner receives Life Membership of the Kalgoorlie-Boulder Aero Club at a presentation in the Rotary Club of Kalgoorlie in 1981 only four weeks before his untimely and tragic death. (Photo Joan MacLeod)

He reminded me that during my last check flight, he had discussed electrical failures with me. A proper scan of the cockpit instruments would have revealed a discharge on the centre reading ammeter. The alternator switch should have been placed in the 'off' position briefly, while the circuit breakers are checked. If non had blown, then I should have reset the alternator switch to the 'on' position to re-excite the system, which should then have functioned normally. However, if the electrical power had not been restored, then the electrical load should have been reduced. This was done by switching off all un-necessary electrical equipment to the minimum possible. By doing this, I gained the maximum benefit from the aircraft battery before it expired. With all electrical power lost, I could not operate the wing flaps, operate the radio or electronic

The Kalgoolie-Boulder Aero Club (KBAC) in 1974. Two club aircraft visible include VH-PPL a Piper Cherokee PA28/140 (background) and a Victa aerobatic trainer VH-FHP. The other aircraft used was a Cessna C150 registered VH-KPK. (Author)

navigation aids, aircraft lights or certain flight instruments that needed electrical power for gyroscopic operations. I had obviously suffered an alternator failure only fifteen minutes after departure from Kalgoorlie and I was glad that I had asked Kalgoorlie Flight Service to keep a radio schedule with me because if anything nasty had actually happened, at least they would have raised the alarm. Quite a few onlookers had gathered from the club to see what all the excitement was about, which did little for my already slightly damaged ego. The aircraft was towed to the hangar with a maintenance release unserviceability and I retired to a quiet place behind the aero club to re-read the aircraft operations manual!

Memorial in Karrakatta Cemetery, Perth WA

Wings in the West

After I completed my unrestricted licence with the KBAC in 1975, I began to think about the requirements for Commercial flying training. I completed a few endorsements on as many local aircraft as my bank account could afford and set about keeping a lookout for anyone that needed to have an aeroplane ferried somewhere. One of the air operators at Kalgoorlie was Trans-West who flew both Regular Public Transport and charter operations in support of the West Australian mining operations at Leonora, Windarra and Laverton. To say that this part of the west was busy with air traffic would have been an understatement. Pilots jockeyed for positions in the landing and take-off sequences and called each other on the VHF radio to ensure that no-one had a mid-air collision. They also had to call the nearest Flight Service station using the longer-range HF radio to advise when they were beginning their descent and in the circuit for landing.

The intensity of the air traffic in this part of the world was to prove very useful to me because I was able to become involved in some of the flying as a private pilot. While I was studying for my Commercial Pilot Licence I made myself available to the local air operators by paying for my endorsements on as many different types of aircraft as possible in order to help out with ferrying aircraft that were not producing revenue for the operator. This occurred for a number of reasons such as an aircraft stranded by becoming unserviceable. An Engineer had to be flown out to the location to fix the aircraft which then had to be brought back to the main base to be put back on the commercial flight line. I arranged my rosters at work with shift-swaps that were mutually acceptable with other officers so that I could make myself available to fly in the aircraft taking the Engineer out to the other unserviceable aircraft. When he had fixed it, he would fly back with the other pilot, and I brought the empty aircraft back to Kalgoorlie. The operations management of Trans-West increasingly began to contact me to help them out and I happily zipped around the countryside to places like Jandakot, Leonora,

Laverton and Windarra picking up or re-positioning aircraft.

On one occasion, I was sitting at home listening to the Mount Charlotte mine poppet head rattling away just around the corner, when the telephone rang. It was the Trans-West office in Maritana Mall inquiring if I was available to ferry a Cessna 205 to Jandakot from Kalgoorlie? I knew what Cessna 182's, 206's and 210's looked like, but a Cessna 205? I figured my C210 endorsement would cover it, and tongue in cheek said that I would do the flight. I rang Arthur at the airport to confirm the legalities of the matter and he assured me that my endorsements were appropriate and offered to give me a check circuit in the C205 if I was unhappy. I drove out to the airport and picked up the flight manual and handling notes for the aircraft. I realised that there was little difference between this aircraft and the C206 Stationair that I had already flown earlier in the week and was quite happy that the C205 would not present me with any problems.

That night, I sat down and carefully made out a flight plan ready to submit at the Flight Service Office in the morning. A weather front had crossed over the state, leaving some unpleasant remnants and I stood in the office next morning studying the weather forecasts and asking the local Trans-West pilots who were also planning for the day for their opinions. They were adamant that I could proceed safely, even with my low experience and laughingly commented that if I was going to be a success as a future Commercial Pilot, that I had better "get into it".

I called Flight Service on the VHF radio and reported taxiing for runway 28 and after take-off reported my departure time to them and climbed to my planned level of 6000 feet. The route followed the direct track to Southern Cross and then on to Cunderdin and Jandakot. All went well until I approached the Cunderdin airfield when the weather seemed to deteriorate quite rapidly. There had been no forecast amendments or information and unhappily I was forced to descend below the murky stratiform layers of cloud that confronted me. I could see the Cunderdin township and its airfield quite clearly below, but looking ahead, there was a lot of low cumulus "scud" around that soon forced me to fly at a lower altitude. I radioed my position to Flight Service but received no reply. I checked my documents to check that I was transmitting on the correct Flight Service frequency but could still not get a reply. My flight plan was lodged on full reporting procedures, which required me to report at the positions nominated on the flight plan form submitted earlier. I knew that failure to report would result in the implementation of search and rescue procedures. There was little I could do. I had not been radar identified and radar identification was unlikely now that I was flying low level behind the hills. Transponders were not used for light aircraft in those days and most Australian air traffic control radars did not have the facilities to cope with that sort of technology or procedures. I looked at my documentation again in a somewhat forlorn mood to check the correct procedures for a radio failure emergency. The checklist told me to transmit my position and intentions "in the blind" and land at the nearest suitable airfield. I looked behind me and was horrified to see a large squall had developed. Cunderdin no longer seemed to be a "suitable airfield". I tried to tune in the Jandakot NDB radio-navigation aid without any success. Similarly, I tried Perth airport and then the Guildford NDB's again without success. I had started a dead reckoning navigation plot on my WAC chart based on the forecast winds, and stared with a sinking feeling about my survival, at the darkened landscape below looking and hoping for a visual clue. I decided to backtrack from the Cunderdin NDB and successfully tuned in the beacon. The station ident "C-U-N" came through loud and clear. The ADF (Automatic Direction Finder) needle swung around pointing to a relative position to the rear of the aircraft. This was very good and I quickly intercepted the correct flight planned track. I just hoped that the ADF equipment was calibrated properly and that I was not leading myself into another debacle.

    Those of you who have flown across the Darling Ranges to Perth or Jandakot under any sort of cloud overcast will know that visual navigation on a good day is difficult. The huge expanse of forests offers few visual clues, unless you are low enough and have a visibility sufficient to see Mount Dale or Mount Bakewell sticking up out of the murky conditions. If you happen to be at five hundred feet above ground level, you may even miss seeing the reservoirs located in the folds of the ranges. However, if you make it to the western foothills, you will easily see Forrestdale Lake and a safe arrival is guaranteed. WRONG! Five miles southwest of Cunderdin, I noticed that the ADF needle was no longer pointing to the rear of the aircraft. I had not changed heading and quickly checked the DG (Directional Gyro standby compass) to ensure my compass settings were correct. I flew a rate one turn to the left for thirty degrees and then back to the right to re intercept the track. There was nothing wrong with the compasses. The ADF needle had settled on a bearing to the left of the aircraft in a position where it could not in reality be. I checked the aircraft avionics and realised that the only explanation was an in-flight radio and nav-aid failure. The fuses and circuit breakers were all serviceable and I began to feel very unhappy. Maybe I was becoming paranoid but I could also sense an odd vibration in the airframe. I scanned around the aircraft looking for the problem and minutes later the engine began to surge gently. The manifold pressure gauge was fluctuating but had not reduced and I therefore discounted icing as a possible cause. In any case this was a petrol-injected engine which did not use carburettors. I broke out in a cold sweat and scanned around the aircraft, frantically looking for a forced landing area but there was nothing suitable. If the engine failed now, I would most certainly finish up in the trees below! On the western side of the hills I could now just see the plains which appeared as a bright ribbon of light between the heavy overcast cloud conditions and the darkened landscape below. I reached down to tighten the throttle friction nut but remembered the frequent warnings from Arthur about tampering unnecessarily with the controls. Clearly something was wrong with the engine, but the facts were simply that the aircraft was flying within engine and airframe limits. If I tampered unnecessarily with the power, I may well induce a power failure. I sat very alert and very concerned. I tightened my harness and stared ahead miserably hoping that nothing bad was about to happen. I switched the radio over to the Jandakot Tower frequency and listened carefully. No one seemed to be transmitting and so I made another call on the VHF radio hoping that I would be heard. I was astounded when Jandakot Tower answered. The controller immediately berated me for not reporting my position at Cunderdin and asked me for my position. I stared around the aircraft. I only had a Dead Reckoning position and was feeling exhausted from the stressful situation I had found myself enduring. I made an assessment based on the plot I had drawn on my map and asserted that I was about 8 miles to the east of the airport. He advised me to overfly the airfield and report overhead. I maintained my heading and suddenly saw the airfield to my left at about two miles. I turned gently to establish myself overhead and after reporting was told to enter the circuit on a left downwind position and report on base leg. My brain began to work again and I conducted my pre-landing checks. The aircraft did not feel right and the engine was behaving in a most odd manner occasionally vibrating and shuddering. The Manifold Pressure gauge was fluctuating in a most erratic manner and I started to wonder if there was a possibility that I had suffered a fuel leak and was about to run out of gas. I reported on the base-leg and was cleared to land. As I turned onto final approach to land, the engine suddenly surged and ran flat-out. The nose reared up in front of me in a savage climb and I grabbed the throttle and tried to pull it to idle. The throttle did not respond and I had to use the trim to get the nose down again. The runway was still ahead despite the sudden and unexpected climb. I assessed that I was far enough from the runway threshold to glide-in safely without power and pulled the mixture control to "off" to cut off the fuel supply. The control tower started transmitting in the background asking me to confirm that operations were normal! I replied quickly that I had a stuck throttle and was conducting a glide approach. He asked if I was declaring an emergency but as he did so the engine stopped quite suddenly with the propeller windmilling and I dropped the microphone. I remembered the old airman's adage to aviate, navigate and communicate in that order. I was already struggling with the controls and then discovered that the flaps were not working, so I could not slow down. I was getting close to the runway and there was no time for the radio as I gently carried out a sideslipping manoeuvre by crossing the ailerons against the rudder. Sweat from fright was now pouring off me as the runway rushed up towards me in the closing stages of the landing. I floated a long way down it as a result of the excessive speed, which I had failed to minimise in the panic. I touched down reasonably well and used the last vestiges of the landing to run off the bitumen across the grass flight strip to the edge, so that I would be well out of the way of other aircraft trying to use the same runway. How I had the presence of mind to even think about other aircraft is still a source of amazement to me. How fresh the air felt as I turned off what was left of the electrical systems and climbed out of the aircraft. A safety vehicle arrived with an engineer on a tug motor and the aircraft was towed to the parking apron. Investigation revealed that the throttle linkage had failed badly and had been the cause of the surging power and misbehaviour of the engine. I had received a nasty fright and had been lucky to get out of the mess unhurt. The lesson I learned from it was to always have a plan that minimised the workload in the cockpit by avoiding situations that could easily get you into trouble. Despite this rather crude lesson I began to look for any opportunity to go flying on other cross country flights as soon as I could, so that I could consolidate my newly acquired flying and navigation skills. As an experienced pilot once said to me, "the pilots who have stopped learning all there is to know about flying, are dead!" Each flying trip was an opportunity to learn something new. One of the jobs I was given was to ferry a commercial pilot from Kalgoorlie to Windarra so that he could pick up an aircraft that had been stranded there by an engine unserviceability and continue on a charter flight from Windarra to Meekatharra. The company engineers had already been up there on an aircraft from elsewhere and rectified the problem.

 The landscape was coloured typically West Australian iron red interspersed by large white salt lakes, stretching inland as far as the eye could see. The visibility was almost crystal clear without a cloud to be seen anywhere. The Cessna 206 was fairly slow at 125 knots as we flew along in the warm outside air. I sat back pleased with myself and was beginning to feel more confident about becoming a commercial pilot. The high wing of this aircraft type afforded an uninterrupted view of everything below and I waited patiently for Yerilla homestead, which I had nominated as a flight plan check and reporting point to come into view. I looked at my flight plan, that showed my estimated time of arrival overhead Yerilla should be 0215 UTC (GMT in those days!). I glanced across at my passenger. He was fast asleep and I briefly admired his uniform with its Trans-West wings on a blue shirt and two gold bars on each shoulder. Soon I would be studying for my CPL and looked forward to the day when I could fly for a living too.

 I took great care with everything and double-checked my flight plan and then checked all around the cockpit to make sure I had not forgotten anything important. I did not want any embarrassment from some foolhardy oversight! I checked my watch. The time was now 0210 UTC and I looked hopefully over the nose of the aircraft searching for Yerilla homestead. With only five minutes to go before passing over the top, Yerilla should have been clearly visible, even if the weather had been poor it should have been visible, but I could not see it anywhere. A little puzzled and with growing concern, I checked my watch again and my departure time from Kalgoorlie. I had flown past and correctly identified the township of Menzies, and this check had shown me to be correctly on track. I had used the Automatic Direction Finder (ADF) on departure, to calculate the wind drift, which I had applied to my track to fly the correct heading. With everything considered, I should still have been on my flight-planned track, but if this was so, then where was Yerilla? My concern was now giving way to mild panic and I began to feel a little hot under the collar. Maybe I should call Kalgoorlie Flight Service and declare an emergency, at least they could help and then my passenger would never know. I checked my headphones and glanced across at my passenger and almost jumped out of my skin as I realised that he was watching me out of one quizzical eye.

"What's the matter mate?" he inquired.

" I can't see Yerilla anywhere, at the moment anyway..." He opened the other eye.

My voice trailed off miserably. I had really hoped to impress this Trans-West pilot, so that he would speak favourably about me back at the office, but my aspirations were fading rapidly.

I stared around at the landscape and then at my watch as the sweep second-hand swept without mercy past 0215 UTC. My passenger was now staring at me, his face expressionless. "What time should we be there?" he asked.

I showed him the flight plan and pointed to the flight-planned estimate for Yerilla of 0215. "Well what's the time now?" he queried. I pointed to my watch, and he quickly checked his own. He resumed staring at me. No good just staring at me, I thought to myself. It would be more to the point if he looked outside and gave this idiot-would-be-commercial pilot a hand in getting unlost! I was starting to feel very silly indeed. He spoke again. "Are you on track?" "I thought I was" I mumbled almost incoherently.

"Well if you are on track and you have reached your estimated time of arrival, Yerilla must be down there, mustn't it?"

I agreed and looked hopefully at the ground and then at my map.

The rules said that pilots should always read from the ground to the map and not the other way around. Or was it?....

My passenger seemed to be very patient about all this.

"Well is it in front of the aircraft?" he asked. I shook my head unhappily.

"Is it behind you?"

I glanced behind the aircraft to the left and then to the right. I waggled the tailplane from side to side, so that I could see behind a little better, but I could not see Yerilla anywhere at all. I was now sweating with embarrassment. My pilot companion seemed amused. How could he be so calm when I appeared to have got us both lost? "Is it to the right of the aircraft?" he asked?

I looked, but knew in my heart that it would not be there.

He continued, "Well is it to the left then?" I looked to the left to try and be convincing, but my hopes of a career as a Commercial Pilot had now definitely evaporated.

He finally exploded. "Well if you are on track and on time." He paused, "Yerilla isn't behind us, or up ahead, or to the left or to the right...is that correct? If so...it must be..."

At this point, he took the controls and positively rolled the wings in a very steep bank to the left. His left hand shot passed my nose waving an accusing finger at the ground underneath the aircraft.

"....underneath the bloody aircraft, wouldn't-cha-reckon?!"

Kalgoorlie Airport 1975. At the top left of the picture is the RFDS hangar. Next door is Marion Aviation with the longer building housing the Kalgoorlie-Boulder Aero Club. At bottom right is the Flight Service Unit with a yellow DCA Ford Falcon parked outside with the green lawns of the Meteorological enclosure directly behind. The newer building to the left of the Flight Service Unit is the Airport Terminal building. The coloured cars belonged to the hire car companies. (Author)

Down below was Yerilla, passing behind the aircraft at a speed close to 125 knots! I was right on the correct track. He roared with laughter and straightened the aircraft on my original heading.

The slant range visibility even in apparently clear weather conditions can still conceal visual checkpoints until you are just about vertically overhead. Humiliated, I sat working out an estimated time for our arrival at Windarra. I couldn't believe my stupidity, especially since my instructor back at the Aero Club had painstakingly explained the same thing to me half a dozen trips ago, while I was still training. I looked around and stared at my passenger in the same way that he had been doing to me. "Another problem?" he asked.

"Nope" I replied, "but if you happen to see Windarra going by shortly then maybe you'd better let me know, or we'll never get down!"

He chuckled and made a rude comment about my mother under his breath, but I had learned this lesson very well and to this day I always have a look under the aircraft, if I cannot find a visual feature. You would be very surprised at how many times this very elementary procedure has saved me in all sorts of weather from getting lost during my career.

Kalgoorlie airport was an active centre for light aircraft charter operations with an increasing number of jet aircraft appearing as mining ventures began. Naturally, with the increased numbers of aircraft operating, the number of emergencies also increased particularly since the vagaries of the diverse weather patterns could catch even the most experienced and careful of pilots. Throughout most of the year the weather conditions were very pleasant with little or no cloud. Sometimes statewide thunderstorm activity or cyclones in the north, wreaked occasional havoc and in wintertime during the northern dry season, fogs were not uncommon. Fogs normally occurred on clear cloudless nights and usually formed in one of three ways. Radiation fog formed when moist air-cooled below its dew point, by contact with a cold land surface, which had lost its heat by radiation into space. Sometimes moist air flowing from relatively warm seas to colder landmasses would cause advection fog to form in coastal locations. Radiation cooling transfers heat from the land into space and assists in the formation and maintenance of this fog. Over the decades, fog has been difficult to predict. It requires delicate atmospheric conditions for its formation, but when it does form, it can often do so with amazing speed.

In 1971, these problems caused an air drama, which involved a passenger jet approaching to land at Derby in the Kimberley region of northwestern Australia. Fog appeared from nowhere and Flight Service and Operational Control featured prominently in saving a situation, which could have become a disaster. The FK28 Fokker Fellowship jet service departed from Port Hedland with approximately fifty passengers on board for a routine flight to Derby. Fog had closed the airport at the coastal resort of Broome, but no fog was forecast at Derby which is located 89 nautical miles further up the coastline.

If the weather forecast had shown marginal conditions, then extra fuel would have been carried to permit a safe diversion to an alternate airfield where the weather conditions were more acceptable. This was and still is a problem for air carriers, because if extra fuel is carried for diversion purposes, then cargo or passengers have to be reduced to prevent overloading, since most aircraft cannot fly with a full fuel load and maximum payload. The reduction of the payload will obviously reduce business revenue and so strict adherence is given to the regulations to maximise operating efficiency and ensure that a safe fuel margin is provided. Under the current circumstances, the weather conditions forecast for Derby indicated that the air was too dry at that location for the formation of fog and did not require the carriage of extra fuel. A new problem started at Port Hedland where fog was actually being forecast and was in fact beginning to form, threatening to prevent the departure of the jet. If it remained on the ground it would be stranded for the night, which was not in the best interests of the passengers. The aircraft therefore departed quickly before the fog closed in the Port Hedland airport. However, while the aircraft was safely en-route, the Flight Service Officer on duty at Derby airport noticed that a mist was beginning to form not far away from the airport and contrary to the current weather forecast that the jet was using.

He knew how quickly fog can form and wasted no time in reporting the matter to the captain of the inbound FK28. The FK28 crew became justifiably alarmed, but as they descended towards Derby they could see the aerodrome light beacon flashing and they could also see the lights of the township and had no reason to believe that they could not make a safe approach and landing. Without any warning, the local weather conditions began to change rapidly and fog began to form at a rate that no one had seen before and the runway literally disappeared from the view of the approaching jet. The captain carried out a normal instrument approach procedure using the ground base radio navigation aids but could not see the runway from the minimum safe altitude and was obliged to execute a standard missed approach procedure and climb back to a safer height again. The fog had now concealed the aerodrome completely leaving the jet crew with a very unpleasant situation, because little fuel remained to divert to a reasonable alternate airfield. The situation worsened at Derby with the Flight Service Unit reporting that the fog was now thicker and was seemingly more widespread. There were now no options. The captain radioed a Mayday call and the FSU alerted the Rescue Coordination Centre in Perth. The Senior Operations Controller (SOC) quickly assessed that the closest airfields suitable for a diversion were out of range. Wyndham was 277 nautical miles further along the coast and Kununurra was 301 nautical miles away. He asked the crew to verify the fuel available and confirmed that the aircraft was seriously faced with a forced landing at night in the bush.

The duty FSO alerted the Derby OIC Mr George Moyle who took little time to race to the airport FSU to try and help with the situation. Flight Service staff have intimate knowledge of the area for which they are responsible and the staff on this occasion established that the only hope for the FK28 was to go to Fitzroy Crossing. The airstrip was not equipped with electric runway lighting but if they alerted the township in time, flares could be laid. The weather there would be no problem and so the matter was suggested to the captain of the aircraft. He and his first officer calculated that enough fuel remained to reach Fitzroy but a return trip to Derby was out of the question. Derby Flight Service advised the Perth Rescue Coordination Centre of the situation and immediately telephoned Fitzroy Crossing. However, the telephone exchange was closed at night and the exchange switched the line each night through to the tiny local hospital in case of emergency. This line was also connected to a local telephone box in the township so that the locals could also call the hospital. The nurse on duty answered the phone but could hear nothing and decided that the constant ringing was caused by local drunks and she therefore ignored it. Eventually, she answered the telephone again to scold them and was horrified when she learned of the approaching emergency. It was all scarcely believable. The word was spread quickly around the township and in typical Australian outback style, the locals galvanised into action to provide all possible assistance with no thought for anything else but for the aircraft and its occupants. The airstrip consisted of unsealed compacted material and the locals worked feverishly to lay a flare path supplemented by car vehicle headlights. There were not enough flares for the job but they were spread out as best as possible. The captain sighted the lights before the strip was ready and radioed that a landing would have to be made without delay because of the critical fuel situation. At this low altitude the FK28 was burning fuel at a furious rate. On final approach to land the captain realised with mounting despair that the flare path had not been laid correctly and that he had already flown along some of the available runway with no chance of stopping before the end. At this low level, radio communications with Derby worsened and a long silence caused everyone there to fear the worst. Meanwhile the embattled captain, forced to overshoot was now doing a very tight 'U' turn to get back quickly and land on the other end of the airstrip where he now knew the lights were correctly laid. This manoeuvre is difficult enough with normal aerodrome approach lighting, but in the pitch black of the West Australian outback, the judgement of the Captain's extraordinary flying ability was proven as he lined the aircraft back onto final approach with absolute precision.

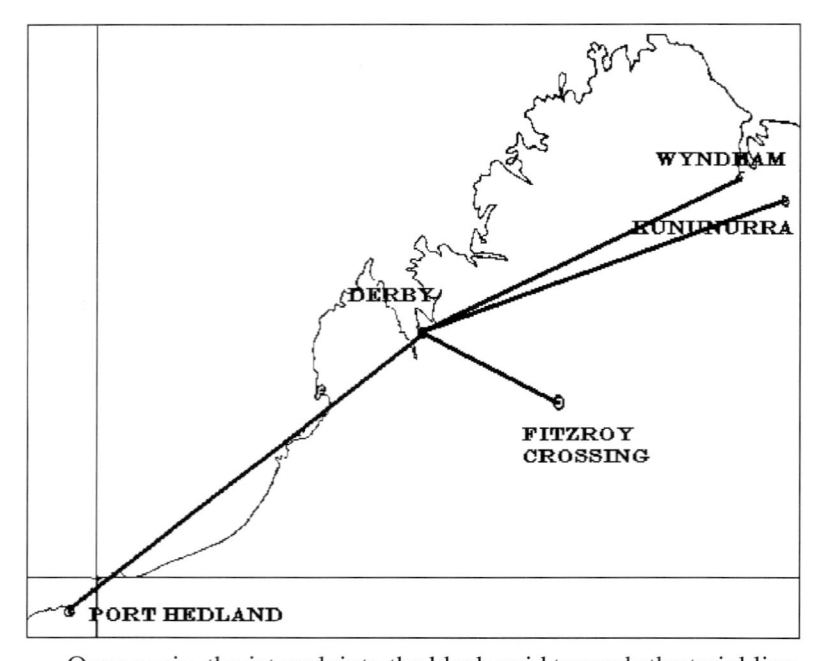

Map showing Wyndham, Kununurra, Derby, Fitzroy Crossing, and Port Hedland with flight routes.

Once again, the jet sank into the black void towards the twinkling flares on the airstrip. The crew must have been numbed by the sight of the fuel gauges indicating that the fuel had expired. Expertly the captain flared the aircraft for landing and touched down safely. The performance of the Captain and First Officer can only be described as brilliant. The local police vehicle manoeuvred in front of the aircraft as it slowed to a stop and indicated for them to follow him to the apron area. A radio call to Derby to report arrival was successful through all the HF static and everyone breathed a very heavy sigh of relief. As the jet began to turn to follow the police vehicle the aircraft engines both died. How the aircraft was recovered from the airstrip later, is another story. Aircrew are now left pondering if satellite communications and microwave links are a good thing, with the loss of the outback units and the years of knowledge they had accumulated about their local area. Only time will tell. Whatever problem occurred, support from all Flight Service Units and their supporting OCC/RCC's was guaranteed.

Incidents like the Fitzroy Crossing diversion caused many officers working in outback Flight Service Units to remain on duty as a matter of habit, without thought of payment long after their rostered hours, to ensure that a lone pilot returning from the wilderness arrived home safely. I often waited back for such flights and passed the time operating the AWA CR6-B HF Receiver (which had a multi band capability from 200 kHz to 25 mHz), to listen to world stations and see which station I could receive that was the longest range from me. For example, I discovered that the 13288 kHz frequency on the South Eastern Asian network 3, (which included Darwin), was also used by Vancouver Aeradio and sat fascinated listening for hours, to aircraft calling in from various parts of the Canadian seaboard.

On Boxing Day morning of 26 December 1974, I was on duty as usual in the Kalgoorlie Flight Service Unit, listening to various network frequencies. I could hear traffic passing position reports to the Katherine and Alice Springs Flight Service Units on the 8938 kHz frequency for locations that I knew were in the Darwin Flight Information Area (FIA). I was very curious, because I had not heard Darwin on the air at all and commented to my colleagues that the Darwin Flight Service Centre must have gone on strike or something. My interest grew rapidly as each aircraft called with a 'Rescue' callsign and seemed to belong to RAAF C130 Hercules transport aircraft and I began to ponder the reasons for this activity. Maybe we were at war? I discounted that, as they were unlikely to be using 'rescue' callsigns, however, I took the liberty of asking other network stations close by, if they knew what was going on, but everyone else seemed as mystified as I was. When the afternoon shift arrived for work they excitedly told us about the morning news which had shown the first reports coming in from Darwin about the damage done by a very nasty cyclonic storm named Tracy. Very soon the most intensive airlift ever known in Australia was in progress with aircraft from the RAAF, TAA, Ansett Airlines, Qantas, USAF, RAF and RNZAF to name but a few, flying shuttles airlifting refugees out to safety. They returned loaded with supplies of medicine, food, water, and emergency equipment for those still stranded there. On one day alone 32 aircraft flew 40 flights and carried nearly 7000 people out of the area. In just six days over 22,000 people had been airlifted from Darwin, while construction crews began the task of rebuilding over 10,000 homes. I was glad that I had left Darwin before that disaster but often wondered what it would be like to live through a real cyclone. Little did I realise that I would have my fair of cyclone experiences later and even fly on an Air Force cyclone penetration flight a few years later!

Fokker Fellowship FK28 jet taking off at Kalgoorlie, similar to the aircraft used in the Derby-Fitzroy Crossing incident in 1971 (Author)

The national economy was reasonably good and there was plenty of work for the air charter companies. In between my duties as a Flight Service officer, I continued to fly in my spare time. I did not always manage to fly the fleet aircraft and on one occasion I was asked to ferry an aircraft to Mount Newman in the far northwest of the state. I was asked to go down to the hangar and see the engineer who would give me the details. When I arrived there I was dismayed to find that the aircraft to be flown was a scruffy and rather forlorn looking Cessna C150 trainer. The pilot who had been flying it had been delayed through bad weather and left it at Kalgoorlie, hoping that someone could pick it up later and return it to the flying school at Newman. The aircraft had seen better days and was not equipped with navigation aids or an HF radio. The VHF radio looked as though it had been stolen from a museum and its reliability was

180

doubtful. None-the-less, the paperwork was all in order and so I submitted a flight plan to Kalgoorlie Flight Service. I was obliged to nominate a SARTIME for arrival at Newman because the lack of an HF radio made en-route communications impossible. I would therefore have to rely on en-route radio broadcasts and ask Meekatharra Flight Service as the responsible unit for Newman to keep my SARTIME in order to watch for my safe arrival. If I failed to report in by the nominated time then search and rescue procedures would start. I set off in rather bleak weather conditions, ahead of a cold frontal system that was crossing the state. The best speed I could hope for was 85 knots and this dictated a refuelling stop at Leonora and Meekatharra. I had to navigate using a World Aeronautical Chart (WAC), a pencil and a prayer! En-route between Meekatharra and Newman I had selected only one navigation fix at a place called Beasley Pool. I thought that this would be easy to see from the air and I would then know my exact position. When I arrived at my estimated time for the Pool, I discovered that the Gascoyne River had flooded badly and engulfed the pool making it completely unusable as an accurate ground position fix. I knew I was crossing the river, but at what point was I actually crossing it? Was I east or west of where Beasley Pool should have been? The Collier Ranges ahead were black in the poor light and offered no clues. The ground below was also darkened and I could not even see the Great Northern Highway. I flew on miserably, not lost, but not quite sure of my exact position. I had been regularly broadcasting my previous positions in the hope that someone had heard me and relayed the information to the nearest FSU. The reason for this was that when an en-route report was received by the FSU on relay from me through another aircraft, the search would start from the last position received from me instead of the departure point if I went missing. The radio was certainly ancient! I worked out a dead reckoning navigation fix and headed hopefully for Newman. Eventually, I spotted the layered ridges of the surface mines near Mt Whaleback at my destination and I headed directly for them. After seven hours of flying time, my navigation skills had improved dramatically and I wearily clambered out of the aircraft at the

Newman airstrip. I quickly found a telephone and made a reverse charge call to the Meekatharra FSU to cancel my SARTIME. They were happy to hear from me and told me that the weather between Meekatharra and Kalgoorlie had really turned nasty. The instructor at the school knew nothing of the aircraft and told me that I had better return it to Kalgoorlie!

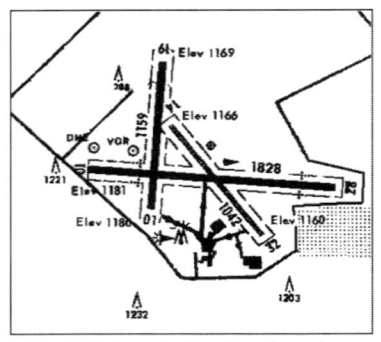

The old Kalgoorlie/Boulder Aerodrome closed down in 1992

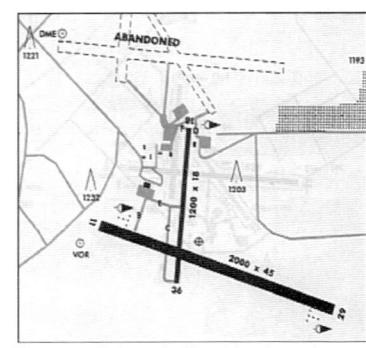

The new Kalgoorlie/Boulder Aerodrome opened in 1992

I had just spent seven hours dodging bad weather and enduring engine fluctuations and using fuel at an alarming rate. I badly wanted the flying hours and fingered the air ticket that I had been given to return to Kalgoorlie via Perth on the jet. I decided to settle for the soft option and suggested the new owner of the aircraft make his own arrangements by telephone. After a long day, I arrived back in Kalgoorlie at midnight for a well-earned sleep. I acted as Officer in Charge at Kalgoorlie airport for temporary periods, while Ken was away and in late 1976 was posted again to eastern states. I was sorry to leave the west and vowed to return some day.

The Kalgoorlie Flight Service Unit was closed on 17 June 1992 at 1900 hrs local time. The Unit teleprinter circuits with the international location indicator 'APKGYSYX' were closed forever on the following day at 18 June 1992 exactly 24 hours later at 1900 hrs local time. Sadly, the Kalgoorlie-Boulder airfield with its long history, was also closed on 14 July 1992 to make way for new housing. The new airport is located only 1.1 kilometres to the south-east of the old site.

182

North Queensland

North Queensland generally encompasses the area north of a line drawn roughly from Mackay to Mount Isa although some parochial local's further south will tell you differently. This area is still largely unsettled and quite remote from anywhere else, connected only by two main highways. The Bruce Highway runs along the east-coast and connects the Queensland State Capital Brisbane with Bundaberg, Rockhampton, Mackay, Townsville and Cairns. The Flinders Highway connects Townsville with Cloncurry and Mount Isa and then links up with the Barkly Highway across the state border into the Northern Territory. By now I had a young family of my own and was working all hours of the day and night. My social life revolved around others working in the same environment because attendance at regular weekly club activities could not be guaranteed, although I did manage to play cricket on a regular basis for one year! The demands of navigation in what is known as the outback of Australia can be very difficult irrespective of whether you are on the ground or in the air. Quite often, settlements have grown up almost overnight and you may fly over a large populated area that does not feature on the most updated of maps. A great deal of care and discipline is required to avoid believing that you are somewhere quite different, even though your navigation data and logic, points to the contrary. Many careless navigators have become lost because they have not kept up a running plot and accurately computed revised headings to be flown and ground-speed variations. Navigational ground radio beacons are not always available unless you are on a route commonly used by a regular airline service. If you are departing from a major centre that has an airport equipped with radio navigation aids and your aircraft has equipment on board that is capable of receiving those ground radio beacons, then the ability to use them will erode as the aircraft is flown further away. Navigation into the interior of the country where such radio beacons are scarce or non-existent relies on seeing identifiable ground features.

This means that weather conditions in these areas must be suitable for an aircraft to fly safely along the last route segment with continuous visual reference to the ground and underneath the clouds. If cloud does obscure the area around the destination and the aircraft is on top of cloud and unable to see the ground, it may descend to a pre-calculated lowest safe altitude. This altitude is calculated by adding one thousand feet of altitude to the height of the highest natural or man-made obstruction located above sea level within a defined area. For example if the highest hill is 3000 feet high and has a 100 foot radio-mast on its summit, then 1000ft must be added for safety which totals 4100ft. This then, is the lowest altitude to which the aircraft may descend if the ground is obscured by cloud. If the cloud base is 4200 feet above sea level then the aircraft can descend safely below the cloud. This is because the cloud base is 100 feet higher than the lowest permissible safe altitude and the pilot will be able to see the ground. If however, the cloud is at 4000 feet then the aircraft will be flying in the cloud at the lowest possible safe altitude 100 feet above the cloud base. In this situation, the pilot has no option but to divert to a safe alternate airfield and land there.

Charter pilots soon get to know the areas they service and the weather conditions that occur there. Weather forecasts and local weather reports must be analysed to obtain the best overall picture of the weather conditions likely to affect a flight. On routes promulgated by the aviation authorities the lowest safe altitudes are already calculated and shown on radio navigation charts. If the route you are flying is not promulgated then all lowest safe altitudes should be calculated before departure. All diversion contingencies must be catered for, particularly with regard to the amount of fuel carried, to ensure the aircraft does not run out of fuel when a diversion is required. Some irresponsible clients insist that the flight must go on whatever the circumstances and businessmen will become unreasonable because they see a business deal as far more important than the problems facing a charter pilot! Such customers have pressured many pilots into making ill-founded decisions, for fear of losing a customer to an opposition company. We have read of the results in the news media when the aircraft has been destroyed in a storm or crashed somewhere on route because judgements were clouded by is known in the industry as "press-on-itis."

184

Layer of Cumulus clouds en-route. The ground can be seen reasonably well and descent below the cloud base can be conducted safely. Photo Author

On one of my early charter flights, I was required to fly a visual flight to Longreach from Townsville with one passenger. This, on face value seemed like an easy enough task. However, the only aircraft available for this trip was a single engine Beechcraft Bonanza. Government regulations prohibited flying in cloud in single engine aircraft for hire and reward with fare paying passengers. Unfortunately, thunderstorms had been very active throughout North Queensland all week long and had caused quite extensive flooding in some areas. Today was no exception, I wanted to fly as always, but was forced to suggest to my passenger that he proceed by surface, road or train, or even by boat in the flooded inland areas! The only real option for him was to hire the more expensive twin engine aircraft so that we could legally fly in cloud. However, the twin was too expensive for him and he insisted that we try in the single engine aircraft with the famous last words, "you never know, we may get through OK." I sat down with the flight plan and a map and tried to find a safe way through the coastal mountains and at a height that would keep me clear of cloud. The best course of action seemed to be to fly almost due south past Mt Ellenvale on the end of the Paluma Ranges, to the old wartime airstrip at Reid River. Here, a gap allowed passage through to the township of Charters Towers. The country flattened out considerably from Charters Towers onwards and would not present much of a problem, providing there were no thunderstorms lurking there. I was also aware that all the dirt airstrips in the area were currently unusable because of flooding and could not be used if the weather prevented us from proceeding or even turning back.

My passenger was now becoming very restless and quite impolite and so against my better judgement, I decided to attempt the flight. This error of judgement was to nearly cost me not only my hard-earned licence, but also my life. The continental engine sounded beautiful as I climbed the Bonanza to 1500 feet and flew past the western slopes of Mount Stuart on the outskirts of Townsville. Over the Ross River Reservoir, only ten miles to the south, the cloud forced me down to 500 feet above the ground. My passenger lit a cigarette and slumped back in his seat with a superior smile on his face.

I battled on in the increasing turbulence. The Townsville radar controller transferred me to the Flight Service frequency and I called them with a radio check on the VHF radio. I would soon be out of range of VHF and so they transferred me to an HF frequency so that I could maintain continuous communications throughout the trip. Woodstock airstrip was only five miles further on and I could barely see it in the rain that was now becoming almost torrential. I located the Flinders Highway and followed it to Reid River Township.

Deteriorating weather conditions near Mount Ellenvale near Reid River North Queensland. Descent below these clouds without full visual reference to the ground was inviting disaster. Photo Author

My instincts were beginning to tell me that I should abandon the trip and go home, but the rain eased momentarily showing just a veil of mist ahead with the sun shining through. The trap was set.

With a hopeful heart, I aimed the aircraft at the sunlight between the hills, thinking that I was heading for a gap in the weather. Mount Ellenvale loomed 1700 feet above my right wing when suddenly everything went incredibly dark and I lost visual contact with the ground. The rain fell with such force it sounded like someone was emptying continuous road gravel on the windscreen. I looked across at my passenger. He had fallen fast asleep. I was below the lowest safe altitude, in cloud with no visual reference to anything including the ground. In seconds I was going to smash at a 160 knots into the side of the Paluma Ranges. Fear took over from reason. I slammed the nose skyward in a bid to out-climb the rock filled clouds around me. Sweat oozed down my face as I watched the altimeter climb slowly and waited for the bang that would signify my last breath on this earth. The area safe altitude was 5100 feet. At this height, I would be safely established 1000 feet above all the high terrain in this area. I switched back to the radar-controller's frequency in Townsville and requested a clearance to return via an instrument descent using the Townsville VHF Omni Range and Distance Measuring Equipment radio-navigation beacons (VOR/DME). Now, I had broken all the rules! Feverishly, I set myself up for the electronically controlled instrument descent that would let me come down safely below the level of the hills at Townsville. I carried out the required procedures and landed safely. As I taxied in, I felt very foolish and sheepish and my passenger awoke with a start staring around himself. "See, I told you," he yawned triumphantly. He looked at his watch. "Gee, my watch seems to have stopped, what's the correct time pilot?" I curtly informed him that it was 30 mins after our take-off time and that we had landed back at Townsville. I also commented that he was lucky his watch had not stopped because we had crashed! With a stream of oaths, my passenger scuttled off presumably back to where he had come from and I smugly basked in having the last word. It was short lived. I turned around and walked directly into a government examiner of airmen. He knew what had happened and I received the lecture of my life on my misdemeanours over the last hour or so. He ranted and raved and told me that if he ever saw me breaking the rules like that again, I would be heavily fined and have my pilots licence suspended. I had already had a nasty fright and that must have been obvious to him, because he suddenly strode away shaking his head in disgust, leaving me standing, wishing I could be somewhere else. I was relieved at getting off so lightly, but at least I knew now, that I had learned a valuable lesson and would never ever make that mistake again. I observed my ex-passenger marching over to harass an opposition air charter company and to this day I wonder if he is still alive.

A sense of humour was always needed as a charter pilot. He or she is really an airborne public relations agent. Client confidence has to be fostered and some clients could be quite impossible. Their self-importance is quite intolerable at times and the glamour of flying is lost in the reality of being an aerial taxi driver. The support received from Flight Service, especially when flying in remote areas made life for the air operators easier. Pilots and Flight Service Officers often met socially at local gatherings and a good rapport could be maintained because everyone knew what to expect from each other. It was good to know that help was only a radio call away.

I was once assigned to carry a news-reporter to a small Queensland town so that he could provide news coverage for the funeral of a local well-known criminal. The flight was routine and as we flew across the outback areas we chatted and sweltered in the heat of the day, listening to other aircraft reporting to Townsville Flight Service on the radio. The airstrip was easy to find and after we had landed, I decided to wait there for my passenger to avoid delays. My passenger drove off at high speed in a borrowed utility and I sat under a tree reading a flying magazine. An hour or so later, the peace and quiet of my world exploded. I heard the car coming along the track long before I saw it and a sixth sense told me that something was badly wrong. The vehicle came into view around a bend travelling at a lethal speed, trailing a long cloud of red dust behind it. I ran to the aircraft and readied it for an immediate departure. I climbed in and started the engines. Looking back, I could see a number of other vehicles, obviously in hot pursuit of the first and one of them seemed to be backfiring badly. The first vehicle stopped only feet from the aircraft and I watched in amazement as my reporter passenger jumped out screaming for me to "get going."

I decided to get an explanation later and taxied at a lethal speed out onto the airstrip, while my hysterical sweaty passenger slammed the cabin door shut and strapped himself into his seat. He implored me to hurry and kept casting frequent furtive glances behind the aircraft at the approaching cars. I didn't have time for radio and slammed the throttles forward and gunned the aircraft down the airstrip. I held the aircraft down low, while I whipped the undercarriage up and listened as the wheels thumped into the wells under the wings. One mile away, I converted all the extra airspeed into height by pulling the nose up and climbing at over a thousand feet per minute to a safe cruising level. We both sat in silence, while I settled everything down and attended to my navigation log. I radioed Townsville Flight Service with my departure time and sat back in my seat. My passenger had pulled out a large white handkerchief and was busily mopping his brow. His camera equipment had been literally thrown on the back seats without a care and clearly something had frightened him a lot. I demanded an explanation. He explained that he had moved too close to the funeral and the relatives had suddenly realised that he was not a family member. When they realised he was a reporter, they had pulled out firearms and chased him down the road trying to kill him. I suddenly realised that the backfiring I heard had actually been gunfire! So much for my army training in the recognition of arms fire!. We spent the rest of the trip in silence. He paid for the flight and I was quietly pleased to see him go. I smiled to myself when I momentarily thought that the Flight Service Officer on duty, who had received my apparently routine calls, had no idea of the drama that had really taken place.

Passengers are not the only hazard to be endured in air charter operations. Flying in the outback brings also brings aircraft landing or taking off, into close proximity with a wide range of wild animals as well as other aircraft. At Lyndhurst Homestead, 140 miles west of Townsville, I advised Flight Service that I would have to circle for 15 minutes because a mob of at least 300 kangaroos was grazing on the airstrip. Unfortunately, there was no telephone contact available for the FSU to call the homestead and so I had no option but to wait until local staff on the ground heard my engine overhead and made their way to the airstrip to clear it for my landing. At another homestead located at the Valley of Lagoons some sixty miles to the west of Ingham, I always circled patiently waiting for cattle grazing on the airstrip to be moved off by station staff. Cattle were usually very stubborn and a lot of time was wasted circling waiting for the landing area to be cleared. While these activities were in progress, I radioed Flight Service in Townsville on the area HF frequency to advise that that I was not landing as yet, but operating in the area and that I would have to call every 30 mins with an 'operations normal' call. This procedure afforded me search and rescue protection, because if I failed to report at the next nominated time they would raise the alarm and a search would commence to find me.

    I have pleasant memories indeed of the Valley of Lagoons. It was a delightful outback Homestead that nestled in the hills on the upper reaches of the mighty Burdekin River. The view from the living quarters overlooked numerous picturesque lakes and teemed with wild bird-life of numerous species. The Homestead was built by early settlers and forms a part of well-known North Queensland history. It is still inhabited by the Anderson family who breed fine show cattle and racehorses there. I was lucky enough to be offered lunch on my very first visit and a uniformed maid appeared unexpectedly and served me with morning tea complete with white linen tea-cloths and silver tableware. Unfortunately every visit there by air required a certain amount of extra fuel to be carried so that you did not run out while you were still airborne waiting for the animals to be cleared from the airstrip.

On another flight I was flying the aerial ambulance into Hughenden which is a rural commercial centre on the main highway between Townsville and Mount Isa. The sun was setting and the ground was becoming enveloped in rolls of grey dust, settling after the day. Despite the reduced visibility caused by the dust, I could see the township lights clearly against the brown dusty landscape, which was blackening with lengthening shadows as the sun set in a brilliant orange and red hue. I spotted the airstrip with the runway lighting already turned on by the waiting ambulance officers for my arrival. Proper runway lighting was a luxury in these parts and the twinkling lights beckoned me downwards. I switched on the aircraft landing lights but they reflected back brightly from the thick dust, like car headlights on a foggy night. I quickly switched them off again to improve my forward visibility while landing. The runway zoomed up towards me and I flared the aircraft for landing. As I did so, a mob of kangaroos frightened by my noisy approach and concealed by the darkness, jumped into my path and bounded across the airfield. Just as frightened, I pulled the nose of the aircraft up violently to avoid impact and rammed the throttles forward to full take-off power to abort the landing. A squeal of fright from the normally cool flight-nursing sister in the cabin behind me, told me that the manoeuvre had been extreme. Fortunately, we did not have a patient on board, and nearly became patients ourselves! The waiting ambulance crew observing the situation, raced onto the airstrip with siren wailing and all flashing lights on to try and scare any remaining animals away. I watched as they moved off the airstrip again and flashed their headlights to indicate that it was now reasonably safe to approach and land. I liked having agents on the ground because of the added degree of safety in the event of anything going wrong. Sometimes, if no one was there to meet you, a radio call to Flight Service would be required to see if anyone could be contacted to come out to the airstrip and assist. Some time was wasted holding overhead, while Flight Service valiantly tried to contact the locals via radiotelephone or by using the RFDS network. I was always pleased to talk to Flight Service, especially at night. They always provided a feeling of security, when you were on your own in the middle of nowhere.

The ambulance crew seemed to have enjoyed the opportunity to race up and down the local airstrip and greeted me cheerfully as I stepped out of the aircraft with the nursing sister. They asked her, how the landing had been and she cheekily replied that it had been great until I touched down, insinuating the landing had been heavy! They all had a good laugh at my expense and I decided to deal with her later!

Unhappily, the patient we had come to pick up was seriously ill and her present condition was not good enough to be moved. The doctors wanted to try and stabilise her a little more before attempting to medivac her out to Townsville. I was very tired and had been flying since 8am that morning. I had been given a dispensation by the Department of Aviation relayed via the Operational Control Centre at Townsville to exceed my legal duty time in order to complete the pick up of this patient. I knew the ops officers very well and conned them into ringing my wife to tell her that I was in for a very late night. My nursing companion made the most of the moment to visit a nearby hospital acquaintance and I made myself at home on a hospital corridor seat, waiting there like a relative of some hospital inmate for the next development. At 07.30pm I was ravenously hungry and asked one of the nurses on duty if I could scrounge a cup of tea. She asked when I had last eaten and seemed horrified when I told her in all honesty, that I had not had anything to eat since breakfast early that morning. She disappeared to make the cup of tea and secretly I thought, maybe she would 'knock-up' a sandwich too. I lay down and before I knew it, I had dozed off to sleep. I was awakened by the nurse who presented me with a plate of steaming hot roast beef, potatoes and vegetables. I was astonished. Glancing at my watch, I realised that I had slept for nearly two hours. She vanished again and reappeared with a huge mug of hot tea and a bowl of hot apple pie and custard. She laughed at my amazement and enthusiasm while I made short work of her efforts. What a girl, I thought. I told her if I wasn't already married, I would probably have proposed to her that very night! She 'whacked' me over the head and told me to behave myself and then vanished down the corridor to attend to her duties. She was typical of the outback breed. Nothing was too much trouble. They were very practical and sensed when someone was in need. I loved their earthiness and sense of fair play. Above all, I liked the fact that their tough exterior belied a warm friendliness and caring, that seemed unique to this part of the world. I finished the banquet and as a gesture of thanks for her magnificent cooking, I washed up the dishes and cleaned the kitchen sink before sinking back onto my makeshift bed in the hospital corridor and falling fast asleep. I was awakened by the doctor a few hours later. He was obviously very tired and was now very concerned indeed about his patient. Someone had placed a blanket over me while I was asleep and pushing it aside, I sat up and listened to him. He did not want to move her, but her condition had not stabilised as he had hoped. Time was running out and she would soon deteriorate further. Any further delay could not be tolerated and I was now faced with prospect of flying her to Townsville for specialist surgery. The staff told me that she was a young outback mother who had been injured in some sort of accident and was now suffering from infections of the kidneys and liver. I knew the matter was becoming grave for the woman. I asked the doctor to wait while I telephoned Townsville Operations for an appreciation of the current weather conditions. My heart sank. There were tropical thunderstorms everywhere and the weather conditions along the way would be anything but pleasant. I wondered how such a pleasant evening could turn into such a nightmare. A quick conference between the hospital staff and myself determined that we would have to make a mercy-dash with the woman despite the weather and her condition. Declaration of a mercy flight was a serious matter but there was no doubt now that this was now a matter of life and death. Air Traffic Control procedures could be abbreviated and as a result I came under the direct control of the Department of Aviation, Senior Operations Controller in Townsville. The Townsville hospital was notified to expect the emergency and I calculated that the flight would take just over one hour. My flight nurse materialised from nowhere and we raced out to the airstrip to prepare the aircraft for take off. The ambulance arrived with a police escort. While we were loading the patient, I noticed a man with three young children standing forlornly at a discreet distance from us. They were accompanied by an elderly lady who seemed quite distressed. The patient rolled her eyes sideways trying to get a glimpse of the group. I instantly realised that they were her family and ran over to ask them over to the aircraft to say proper farewells. They had seen the signs along the side of the airstrip that imposed heavy fines for venturing onto the aircraft movement area and thought they would be fined if they ventured near the aircraft. I was saddened that such things could have prevented them for saying goodbye maybe for the last time in their lives. The young patient seemed to rally as the children clambered over her and her husband held her hand tightly while tears streamed down his weather beaten face. I arranged with the ambulance crew to ensure that everyone was clear of the propellers when I started the engines. With a lump in my throat, I watched the sad little group retreat again to the safety of the fence where they waved goodbye. The ambulance again drove up and down the airstrip clearing kangaroo's out of the way. I tried to contact Flight Service on the area HF frequency but the crackle and static from the nearby storms made reception impossible. I transmitted a radio call "in-the-blind" fervently hoping someone would hear my signal. We roared down the airstrip and began a very gentle turn towards Townsville. Sometime later I established contact with Brisbane Flight Service on 8938 mHz and they relayed my departure to the Townsville Flight Service Unit. I tested every altitude up to 10000feet to see where I would obtain the smoothest ride for the patient. I settled for a cruising altitude of 7000 feet and levelled off warily, watching the lightning that flashed all around us. I could hear the nursing sister reassuring the patient, but eighty miles from Townsville all hell broke loose as we approached the Great Dividing Ranges. Violent updraughts and downdraughts from the worsening weather tossed us around without mercy. I decided not to maintain 7000feet, preferring instead to let the aircraft rise and descend at will, to avoid overstressing the airframe and controls. It made it easier for the passengers too and I informed Flight Service of our predicament. They advised me that there was no one else flying in the area and that I could expect an Air Traffic Control clearance at any level I wanted. The poor little nursing sister worked like a demented tarantula spider trying to sustain our very ill patient. The turbulence worsened and equipment started to shake and rattle violently. Suddenly the drip attachment hanging from the cabin ceiling broke loose and fell on the nurse. Grabbing it quickly, she held it up by hand so that the contents continued to be supplied to the patient. She nursed the patient with her left hand all the time holding up the heavy drip attachment with her right hand. She held it up for thirty minutes without complaint when I spotted the main runway at Townsville and made a straight-in approach for a priority landing. As I landed, the storms closed in behind us like the closing of the biblical Red Sea and torrential rain hammered down on the aircraft. The noise was indescribable. I taxied quickly to the end of the Townsville air terminal where thankfully, an ambulance and a police escort were waiting for the last leg to the hospital. The patient though very weak, struggled to offer us a whispered thanks before being loaded into the ambulance and setting off with the police escort at top speed towards the hospital. The rain never ceased and my clothes were soaked to the skin. I turned to see the brave little flight nurse staring after the departing ambulance. She looked like a drowned rat. The tension of the previous hours seemed to wash away in the rain and we both burst into laughter at the sight of each other. She had 'pins and needles' in her right arm, caused by holding up the drip bottle for so long in the aircraft cabin. We headed to the deserted air terminal in search of a cup of coffee.

The Author at the controls of the Partenavia PN68 aircraft used for medical evacuations in Townsville. Photograph taken in 1986. (Author)

It was nearly four o'clock in the morning and she had not made one complaint during the whole trip. I never found out what happened to that patient but flying the air ambulance was very rewarding and I often wished that I could fly for the RFDS on a permanent basis, but for the moment I would have to settle for health department contract flying.

Charter flying is very demanding and sometimes not everything goes according to plan. I was rostered to fly the daily air service to Palm Island from Townsville. The twin engine Beechcraft Baron BE58 was my favourite aircraft and I was looking forward to the eight flights planned to the island for the day. After filing a flight plan at the Townsville airport briefing office, I headed to the tarmac to check the aircraft loading. I was intercepted by the Chief Pilot who told me that the Baron was unserviceable and that I had been re-programmed to fly a single engine Cessna 206 borrowed from another company. I was disappointed, but the 206 was a great work-horse and ideally suited for bush and island types of operations, but it lacked the speed and responsive handling of the BE58. The aboriginal passengers had already arrived and waited while I rushed about, refiling the flight plan for the slower aircraft and adjusting fuel figures. My passengers usually travelled in the twin engine aircraft and were not too pleased when they discovered that they were about to fly across 33 miles of ocean, in a single engine aircraft. The thought was made worse by the fact that the Cessna wing is above the cabin rather than under it and if a ditching in the sea occurred, then the aircraft would most likely float on the wings with the occupants trapped underneath.

    The Townsville local temperature had already soared to 34 degrees Celsius and my patience was beginning to wane. We all clambered aboard and I taxied out to the runway holding point for a take off on the northward-facing runway, runway 01. I completed the pre-take off checks and was cleared for take-off. We became airborne and I completed the automatic transfer to the radar control frequency 126.8 mHz. The radar controller identified my radar blip and I completed the 'after take off' checks. The undercarriage on this aircraft was not the retractable type and so all I had to do was raise the flaps, set the manifold pressure to 25 inches, set the RPM to 2500 for maximum climb power. Then I had to ensure the fuel tanks were feeding from the correct tank and switch the fuel booster pump off. Something stirred a warning in my brain. I glanced down at the fuel cock on the floor again. FUEL! Incredibly, the fuel cock was in the "OFF" position. I always tried to be very careful with all aspects of my flying and I could not believe I had got this far with the cocks off. I knew an engine failure was imminent. Angrily, I slammed the cock around to face the other way on the 'both tanks' position. Extremely irritated at myself, I switched the fuel pump on for safety to ensure there was no possibility of air in the fuel lines and sat back to resume my navigation. The Townsville radar site on Cape Pallarenda disappeared under my left wing and I looked ahead at the water and the peaks of the hills on Palm Island in the distance. Thirty seconds later the engine stopped completely. The nose pitched downward galvanising me into immediate emergency checks. Fuel was on, the fuel tanks indicated they had plenty of gas in them, oil pressure and temperature was OK and I flicked the fuel booster pumps on. The eerie silence continued. I checked the fuel mixture control was in the 'full rich,' position, checked the magneto switches, and trimmed the flight controls for the best glide speed. I was about to radio a 'Mayday' call, when the fuel cock caught my eye again. I think my guardian angel was teaching me a lesson today. The fuel cock was in the 'OFF' position. I rarely fly the C206 and the Cessna types of aircraft I usually flew, had the fuel cock assembly rigged in the opposite direction. Instead of reading the 'OFF' and 'ON' labels, I had simply positioned the lever by force of habit. The aircraft was still descending. I told Air Traffic Control that I had left my assigned level with a small problem but would be OK shortly. Angrily I banged the fuel cock around to the original position, closed the throttle and restarted the engine carefully re- applying the power to avoid overspeeding the propeller. The incident was all over in about two minutes, but it seemed like a lifetime of stupidity. I climbed back to regain the 300 feet of altitude that I had lost and resumed my cruising level. Feeling very silly, I glanced back at my five aboriginal passengers who were sitting wide-eyed and silent.

The dry season in North Queensland produced very pleasant flying conditions and on another flight, the day was no exception. The morning was almost crystal clear and cold enough to wear a warm jacket. But the daytime temperatures did not take long to rise and my aircraft, climbing over the Paluma Ranges, north of Townsville was performing very nicely in the smooth conditions. The trip for this morning was a charter flight to the Kidston gold mine some miles to the north of the township of Hughenden. A smoke and dust haze in the distance reduced the visibility, but I could see the Greenvale-Yabulu railway line to the south of my track. Visual navigation can be difficult in these outback areas especially when dusty conditions prevail and so I normally prepared my charts with each 20 nautical miles, marked out along the way. In flight, I calculated my ground speed at my cruising level as soon as possible and then applied the estimated elapsed time expected, to each of the twenty-mile marks on the map.

Today, my ground-speed was 155 knots and I calculated that each twenty miles would take nearly eight minutes to complete and so I annotated the expected estimated times of arrival at each twenty mile point on the map. In this way if visibility restricted my opportunities for finding ground reference points, I always had a good idea of my position. This was called a dead reckoning position. It was particularly useful if Flight Service requested your current position in order to ascertain if traffic conflictions existed in the non-controlled airspace for which they were responsible. The Kidston gold mine was still in its infancy and was only serviced by a dirt airstrip that was about 2800 feet in length. These days some 15 years later, visitors to Kidston will find a proper airport with a sealed bitumen runway, runway lighting, wind socks and radio navigation beacons. But on this trip, I had to contend with navigating visually and hoped that the smoke haze would not prevent a safe approach and landing.

En-route I passed over Zigzag Creek homestead and the Blue Range Mountains which stand silently at a little higher than 2400 feet above sea level. I was on track and one hundred miles from Townsville. The railway from Townsville to the Greenvale nickel mine site had appeared on my left and slowly ran towards me until I passed overhead the position where it meets the mine. I confirmed my ground speed and computed that only twenty minutes remained until my arrival at the destination. I should have passed over the Kennedy Highway which runs between Mount Garnet to the north of me and Hughenden to the south of my position. The dust haze precluded any visually sighting of the highway and that was annoying, since I was going to use it as a check point for a mandatory inbound radio call required at 20 miles from the airfield. I had no option but to continue with dead-reckoning navigation and my estimated time of arrival was eleven o'clock local time. At my current ground speed, I estimated that I would be twenty miles from Kidston at 1051 local time. Kidston was only a small operation during this period and therefore quite difficult to see until reasonably close to the area. The Kidston airstrip is aligned roughly north/south and runs along the western side of the Copperfield River and I flew overhead exactly at my estimated time of arrival. As I entered the traffic pattern, I completed my pre-landing checks and looked around for vehicle dust trails or local smoke that could give me an idea about the wind direction. No such signs existed this morning and I had to consult my weather forecast, which indicated that the main airstream over the state was from the southeast. I continued for a landing from the north of the airfield. The weather forecast agreed with the wind that I computed at my cruising altitude and I was happy that the correct runway for landing has been selected. The Piper Aztec PA23D was a steady aircraft to fly and I enjoyed the solid feel of the controls as I turned the aircraft onto its final approach path. I called Townsville Flight Service on the area HF frequency and reported a safe arrival. This was a mandatory call known as "cancelling sarwatch". This search and rescue watch is based on the time last given as an estimated time of arrival. Failure to make the mandatory arrival call would automatically start search and rescue procedures. The aircraft touched down with a slight bounce on the airstrip and I headed for the wooden access gate at the far end of the landing run. As I bumped across the rough surface, a movement to my left caught my eye. Ambling across the airstrip towards me was a large mean looking Brahman bull. I could see that it had come from a herd of Brahman cattle in the paddock nearby and had somehow strayed onto the airstrip. As I taxied along watching the bull, it began to paw the ground with its left hoof. Menacingly it lowered its head and charged directly towards me. I stared in horror for a few seconds realising that if it hit me the damage could start a fire. The engines were still hot and could easily ignite any fuel from ruptured fuel tanks. Quickly, I pushed the throttles fully open and rapidly turned the aircraft away, back down the airstrip. I fervently hoped that the noise would deter the bull's charge. I stopped momentarily to see if it was still coming and my heart sank. The bull had nearly caught me. Quickly, I pushed the throttles open again and accelerated back down the airstrip, giving some thought to trying a takeoff. I quickly dismissed the idea because there was not enough room to get airborne safely, especially with a tail wind. At the far end, the red dirt was quite loose and the propellers running at full power started a mini dust storm. Holding the aircraft on the brakes, I waited hoping the bull could not see me through dust. The passing seconds seemed like an eternity, waiting for the bang. Briefly, I considered a Mayday call to Townsville Flight Service on the HF radio. I thought that maybe they may be able to contact the local miners to come out and distract the bull. My engine instruments were showing a clear indication that they were overheating. The situation was not doing much for the aircraft brakes either. I returned the engines to idle power and cautiously looked behind for the bull and was very relieved to see him covered in red dust, lumbering away into the bush on the edge of the airstrip. Slowly, I returned to the access gate again. The noise of the engines revving had certainly attracted some attention. Several miners had arrived from the mine to see what was going on. A couple of them had seen the hole in the fence, where the bull first came through and were already busy, repairing it to prevent further incidents. I shut down the engines and with a rather dry mouth from fright and dust, I climbed out into the fresh air. One of miners ambled over with a wicked grin on his face. "Gee mate, I thought you fly boys were all experts in bulldust...what was the problem?" With a roar of laughter he shuffled off to his mates. I managed a weak smile and hastily made ready for the return flight to Townsville. I think that Flight Service would have been rather amused by the nature of my Mayday call if I had actually gone ahead and transmitted it! Even though I did not call, it was nice to have the option available.

    I fully understood the system and its workings and was more than comfortable in this environment. The communications umbrella provided over Australia was very good, considering the resources available. Pilots lost in the outback could always relay on Flight Service and Operational Control for a helping hand with navigation. For most, the simple act of talking to someone on the radio about an in-flight problem was calming. There was no substitute for careful flight planning and adherence to rules.

However, on one occasion, I was caught out rather badly. I was on a flight that took me over the top of the Kidston airstrip southward to Hughenden. As usual, in this area, I was out of range of any ground radio beacons and was therefore navigating visually. At Kidston, I turned onto my flight planned heading of 232 degrees to go to Hughenden. If I had paused to orientate myself, I would have realised that a heading of 232 was not going to point me in the direction of southeast! However, I turned onto 232 degrees overhead Kidston but some thirty miles further on, the details on the ground, in no way resembled the features on my map! Quickly, I measured the track from Kidston to Hughenden on my map. It read 169 degrees, an error of 53 degrees! I swiftly worked out a correction and took up an intercept heading that would take me to Hughenden. As I waited for the Hughenden radio navigation beacons to come into range, I sat mystified about how I could have made such an error. I racked my brains, until finally I remembered that while I was at the Townsville briefing office, someone had asked me for the correct telephone extension for the airport technical staff. The extension was 232! I had been writing the Kidston to Hughenden part of the flight plan at the time and absent-mindedly written the telephone extension in the heading box on the form! Fortunately, I had been in visual contact with ground when the incident occurred, but if I had been flying in cloud, I would have flown myself to a point near Julia Creek, nowhere near where I wanted to go. Many pilots have become unsure of their position and requested assistance and help is easy if radar is available. If they were operating near a capital airport radar facility, the matter is simple. The radar controllers identify the aircraft in a variety of ways, and once correctly identified, can continue to provide radar vectoring to a safe location. However in the outback, there are no such luxuries. A call for assistance results in the Rescue Coordination Centre being alerted and officers there having to manually draw a manual plot of the flight on several charts. The pilot was then asked to advise headings flown, (or not flown in certain instances) and any other information regarding obvious ground features. Prevailing wind

202

information was obtained from the weather office and applied to the navigation plot. When the plot was completed a Most Probable Position (MPP) was deduced. A circle of uncertainty was then drawn around the MPP and an examination of all the ground features in the circle was made. This information was relayed to the pilot in an attempt to assist him or her re-locate their present position. Once this had been achieved, a heading could be computed to the nearest diversion airfield, so that the aircraft could land at the earliest opportunity and the pilot permitted to have a nervous breakdown! Very often outposts were contacted on radiotelephones or via the flying doctor service and asked to listen for engine noises or reports of aircraft sightings. Requests for assistance were also broadcasted on local commercial radio stations and people could telephone the emergency number if they thought they had sighted the missing aircraft. An intelligence officer in the Rescue Coordination Centre then sifted through all the reports and determined which of them could be discarded and which were most likely to be the aircraft. Sometimes at night, if a township was sighted, the town lighting could be momentarily switched off and on at the powerhouse so that the pilot could verify his position. Local residents seemed to be more than obliging in this regard! Airstrips could also be illuminated with flares along the edges of the airstrip so that if the pilot was nearby he could sight them and land as long as the airstrip was long enough for the size of the aircraft. The old adage "where there's a will there's a way", was more than appropriate in these circumstances. If the errant pilot did find a local airstrip at night, his arrival would cause him to become a local celebrity at the local pub for the evening. The bar bill could prove costly, especially for teetotal pilots, but for the locals any excuse would do for a party in places where there was rarely any excitement. Such were my first experiences as a charter-pilot, flying in an era where the Flight Service Units provided the safest traffic information and search and rescue alerting services in the world. Sadly politicians have seen fit to de-commission these units as we move into new ideologies.

My Moment of Destiny

"The people and the events in your life are there. That's the way life is. Only destiny or God can change that. Today my thoughts wander, as I float in a crystal clear sky on yet another flight to nowhere and back. I love this gift that enables me to defy gravity in areas so remote and cruise amongst the atmospheric forces that are so powerful, that I am merely an atom. But today they will try to seize and destroy me as I approach my island destination near the North Queensland coast. I unwittingly fly into the trap, unaware that today, I will brush the face of death.

Freedom of the sky means an escape from the artificial ground smudges that represent a million breathing people. They dwell there defacing the natural folds in the ancient Australian landscape and poisoning the life waters of the land with their material greed and mis-management. But piously, I can throw my head back and laugh at them as I gaze at the spectacle below me and ponder the reality. Up here, I float in a nothingness so powerful, it fuels and changes our daily human emotions. Its invisible atmospheric powers lure me to my moment of destiny. The challenge begins as I sink earthward. The living air around me, tired of my assault, seizes upon this opportunity and boils its wrath behind the hills before reaching the summit and seeking me out. I witness the event in slow motion. My feeble flight controls have no effect and I am forced downward.

Outside, the reality rushes by at over a hundred miles per hour. Inside, as fear is born, a thousand visions from my past, appear and just as quickly vanish. My fight against the elements is lost and crystallises in the despair of my soul, screaming in my brain that I may die. The atmospheric forces rush downward from the mountain, its awesome power too strong for my minuscule existence and I now know that I am doomed to crash. My brain fights for coherence and I shout in despair while carrying out pre-impact checks. Turn off the fuel cocks, bring the power to idle, switch off the electrical systems. There is no time for the radio. I am in the final second and time has stopped. I stare at my tanned hands. *They are beginning to age. They are my father's hands, exactly as I remember them, firm and warm with the veins protruding all the way from the wrists to the finger-tips. Life has been flowing round those veins for the last thirty years. Why must it stop today. My mind is racing with fear. How can I have time for such idiotic thoughts. Death is stalking me.*

I AM GOING TO DIE!

Fence posts, wires and trees form a vanguard and rush at me to do battle. There is a sickening, wrenching, scraping, tearing cacophony, a crescendo of destruction. The fuel tanks surrender to the sudden release of its captive energies and a mist of fuel explodes around me. One tiny spark and I will be no more. My body is hurled forward as the monstrous maelstrom of the crash culminates in impact. I fear the end is close. A mysterious silence envelopes me, interrupted only by the sound of searing hot metal from the now stilled engine rapidly cooling. Life is lingering and I watch my hand reach out as if detached from me. The fingers, seemingly curl, reach for the hatch lever and push it out, allowing the wind to enter, eager to witness the folly of its works. It dances into the cockpit, cooling my face and ruffling my hair. I seem to float from the wreckage and somehow find myself some distance away. There is a still and curious silence. I can feel nothing. The area seems devoid of life except for me. Am I still me? I turn and stare at the wreckage. Perhaps this is death and my body is still in the aircraft. Am I a spirit? I cannot tell. I resist the urge to return and see myself in death. Suddenly, I feel a pain in my back and legs and fall down exhausted. The sun warms me and I gaze upward at the blue sky. A wispy cloud grows, curling and tumbling before growing into a bubbling happiness, drifting lazily across the void that umbrellas our pitiful earthly existence. In moments the cloud has grown and died to be replaced by others frolicking in other parts of the sky. They seem to symbolise the beings that are born here, that live and then die. My mind is a tangle of shock and my thoughts are interrupted by a shout from nearby. Turning, I see two figures running towards me. They are anything but heavenly and sadly, but gladdened, I realise in that instant that I have not died. I have met death in battle and have won. I have experienced a crash. It was my own event. Only me and the elements - and I have survived."

Since that event, I have had a new perspective on flying. The experience described above was how it happened for me at the time. The emotive prose came from the experience and I surprised myself by writing it. Many questions about my feelings are still unanswered but I can attribute most of it I think, to shock. I know more than ever now what my limitations are. Better than that, I know never to exceed the capabilities of the aircraft.

Post-crash pick-up by rescue helicopter crew and Queensland Police. Author second from left. (Author)

I could not see the down-draught of air behind the peaks at Palm Island or see the windshear that caused the dramatic increase in tailwind that nearly caused my demise, but all the symptoms and signs were there. So many things look innocent, yet in a few seconds they can cause more devastation than an equivalent volume of flood-water. In fact I am one of nine or ten pilots who have come to grief near or on the same island including a fatal crash at night in a very sophisticated aircraft. I no longer fly in North Queensland but I have learned some valuable lessons from all of this and constantly question everything that I see. Communication from the ground on this island was always difficult. When I arrived in the circuit area there, the normal procedure was to advise Flight Service to cancel the flight SARWATCH to prevent unnecessary SAR alerting. Now I check and double-check to make sure of my survival and have devised other ways to make sure that someone remains concerned about me until I have safely landed and stepped out of the aircraft. I wish that all other pilots could learn from their mistakes too. Many have asked me about mine and I have not been too proud to tell them. But some will not be as fortunate as I, and we will read about them in the daily news.

Post-Crash Northings

For a while I became quite intolerant of passengers who cared little for air safety and tried to coerce charter pilots to perform tasks that exceeded the performance of the pilot or his aircraft. The threat was always that they would go to another charter flying company if you did not do what they asked and you would lose money. However, after my own crash experience, I was happy enough for this type of customer to go elsewhere. In the long-term, the net effect on my company or me was negligible and so I need not have really worried about limiting our cliental with the possibility of eventually going out of business. There were plenty of people who wanted to fly with a safety compliant air service and bad operators did not last long.

Most of the flights we flew from Townsville in the 1980s were flown to the north and hence I used to say, "I'm off to fly some northings today." Today in the northern dry season of 1985 was such a day and my passengers arrived wanting to carry an impossible payload in our aircraft. The flight invoice said that I was to carry five Telecom employees and their equipment from Townsville to Mornington Island, located in the Gulf of Carpentaria. My passengers and I were to remain overnight there and return to Townsville the following morning. The payload that arrived prior to departure from Townsville, included boxes of dynamite, large sheets of steel, pallets of bricks and heavy machinery. A WWII Douglas C47 military transporter could not have carried that load! The chances of carrying such equipment in a light twin-engine aircraft were absolutely zero. I tactfully tried to explain to the men, that the aircraft could safely carry the pilot, five passengers, their overnight bags and perhaps one or two small toolboxes, but nothing more. In any case the civil Air Navigation Regulations of the day prohibited the carriage of explosives in any aircraft and quite rightly so. Dispensations to the contrary were never given. A heated exchange developed and I spent nearly thirty minutes trying to make them see sense. The men were extremely disgruntled about the affair, but soon made immediate telephone arrangements to have the bulk of the payload sent by express land and sea freight. Their company supervisor decided that I would return the following morning with one of the surveyors, but the remainder would have to wait at Mornington Island for the arrival of their stores and equipment. In the meantime Telecom decided to occupy their workers time by preparing the work sites for the installations required while waiting for their supplies. With the safe load that I had accepted, I still had to off-load fuel to keep the aircraft in a balanced state for flight. I arranged to take on extra fuel at Normanton to make it to Mornington Island after I had burned off sufficient fuel on the way from Townsville to Normanton. The men grumbled and groaned about the situation amongst themselves and one of them, a very large ruddy faced fellow took it upon himself to verbally abuse me at every opportunity. The balance of the aircraft was so sensitive that I loaded the heaviest men in the centre of the cabin and the two lightest men in the rear seats. We took off from Townsville tracking 279 degrees while climbing to a cruising altitude of 8500 feet. The large man continued his tirade of abuse throughout the flight until he incurred the wrath of his workmates who were clearly becoming annoyed with his constant complaints. On the way to Normanton, I began to suffer from drinking too many cups of coffee prior to departure and began to feel an urgent need to go to the toilet. I decided that as soon as we landed at Normanton I would make a swift exit from the aircraft and head for the toilets located in the small air terminal there. Unfortunately, my large abusive companion was blocking the exit from the aircraft with his large bulk.

My needs were now extreme and I decided that after I had landed, I wanted him out of the way as quickly as possible. I turned around in my seat and told him that after we landed at Normanton he was to wait until the propeller outside had stopped turning and then open the door and exit from the aircraft as quickly as he could. I was startled at how cooperative he suddenly became and he stared, I thought, a little nervously out of the window. His baggy eyes widened with anticipation and his jowls quivered with excitement and I mistook this behaviour as sarcasm. We arrived in the circuit at Normanton and flew straight to a base leg in the circuit area before turning onto final of a very short approach to cut down time. I could see no reason why we would not make it safely onto the ground and called Townsville Flight Service to cancel my sarwatch. At this point in the panic about going to the toilet I did not want to run the risk of forgetting to report and then have a major search start for me because I had forgotten to report on the ground. On arrival at the Normanton aircraft parking area, I was pleased with the rapid exit of my abusive passenger from the aircraft and he headed across the airport apron area at an astonishing speed for such an obese fellow. Glad that he was out of the way, I headed straight for the air terminal toilets. While I was there, the large gent appeared behind me demanding to know what was wrong with the aircraft. I was a bit puzzled and asked why he thought there was something wrong? He became quite agitated and insisted that I explain why I had asked him to get out of the aircraft as quickly as possible? I became rather amused and explained that I needed to go to the toilet in a hurry and thanked him for his help. He stomped outside roaring with rage. He had mistakenly thought that there was something wrong with the aircraft, which is why he bolted across the apron area on arrival in case it blew up! His mates exploded with laughter and spent the next few hours endlessly ribbing him. I was personally sorry that I had given him a fright, but I was also pleased that I had finally had the last word! The township of Normanton and the immediate Gulf area on which it is situated is a very interesting topographical and geographical feature. The Norman River runs from the foothills of the Gregory Ranges some 450 kilometres inland to the Gulf of Carpentaria and the fishing town of Karumba. Forty kilometres upstream on the direct track lies Normanton, with a population of approximately 300. The airstrip was constructed from compacted dirt with sealed bitumen ends and was generally useable in most weather conditions. Normanton is a supply centre for the cattle stations and Aboriginal reserves in the Gulf Country and some of the lower Cape York peninsula. The area was served by Regular Transport Services operated by Bush Pilots Airways based in Cairns. This service was taken over by Air Queensland using Douglas DC3 aircraft sporting a distinctive yellow tailplane.

The Normanton area looks very much like the scenes in the popular 1986 film 'Crocodile Dundee' that starred Paul Hogan. The weather is normally hot and humid and the climate generates extremes of long dry seasons and heavy monsoonal wet season conditions that produce flash flooding in a matter of an hour. The Normanton hospital, airport facilities and RFDS have made the township a medical and communications base for a large area comprising mostly of grazing properties. The Normanton centre is connected to Croydon, a gold mining town located 155 kilometres east by a railway track. Both the rails and their sleepers are made of steel, because of continuing extensive attacks made by white ants and other bush termites to wooden sleepers. On my visits to Normanton, I always visited the Burns-Philp store. Burns-Philp was a large Australian trading company with interests in shipping, hotels, retail shops, plantations and travel agencies. Its head office was located in Sydney with many outlets in Australia and offices overseas. The company originated in Townsville in 1873, by two migrants to Australia named James Burns and Robert Philp. I was always fascinated by the range of products available in the business especially in the Normanton store. It consisted of a huge barn style building set high on wooden stumps in typical Australian outback style. One wonders how they coped with termites in view of the problems with the railway line sleepers. The goods available included livestock, bales of hay, motor oil, horse saddles, bridal gowns, vehicle parts, foodstuffs, clothing, plumbing equipment and tools all jumbled about at random. This list gives an idea of the extensive range of items available. I spent hours exploring all the shelves and often found outdated items being sold at bargain prices. The town was quite dusty in the dry season, despite its tropical location. The main street was dominated by the 'Purple Pub' which was the main hotel. Its name came simply from the fact that the exterior facade was painted in a purple colour. I often stayed here on overnight runs and obtained my weather briefings and flight plan information from either Mount Isa or Townsville Flight Service because the Flight Service Units at Normanton and Cloncurry had already been closed down. The Flight Service Officers who were posted to these regions would have found life hard but beneficial in terms of the richness and mateship in the locality that cannot prevail in the capital cities. I found some difficulty believing that the township had been badly flooded in the past, to such an extent that an airlift was organised to evacuate the population.

The wet season, with its summer tropical monsoon rains started early in North Queensland in the summer of 1973. During that time, record rainfall filled the parched watercourses throughout the Gulf country and its environs making crossings impossible. Weather forecasts began showing conditions that presented the local populations and the aviation industry with serious problems. Pilots were legally required to carry extra fuel to cater for en-route diversions when the weather conditions prevented a visual approach and landing. If the extra fuel was carried, then pilots could proceed to a destination that was likely to be closed and try their luck, in case there was a gap in the weather. This was all done on the understanding that extra fuel was carried for the expected diversion to a suitable location where the weather was clear. Usually the alternate aerodromes selected for diversion were to the south of Normanton in the drier and not so tropical atmospheric conditions. By early January in 1974, all roads connecting Normanton with other centres were entirely cut-off by rising floodwaters and shortly afterwards, the only rail link between Normanton and Croydon was also cut. A dirty sea of brown water continued to threaten the town. Soon the township was totally isolated except by air travel. But this option was beginning to erode as the red compacted dirt airstrip became boggy. It had taken over 20mm of rain every day for more than two weeks and this was beyond the normal capability of the best all-weather airstrip. Air traffic had greatly increased at Normanton airport as a result of the road and rail closures.

Some of the aircraft in use included the Douglas DC3 heavy transporters owned by Bush Pilot Airways from Cairns. They ferried supplies in and carried passengers out of the beleaguered town. Fortunately, during this period, the Department of Civil

Air Queensland Douglas DC3 Airliner circa 1980 North Queensland
Photo Author

Aviation was still maintaining a Flight Service Unit there and the major concerns raised about the continued use of the airstrips resulted in careful inspections by the airport groundsmen and Flight Service Officers at least three times each day. These duties were often performed in torrential rain. If the airfield became unusable, then the last avenues for escape would be gone. In the last week of January, all telephone and teleprinter network lines failed, cutting Normanton off from the outside world, except for a low powered RFDS HF transceiver and the Flight Service Unit air and liaison HF frequencies. An emergency meeting of the local Civil Defence committee was called to discuss what should be done. However, while Jim Stack the Flight Service OIC was driving past the airfield on his way to attend the meeting, he was shocked to see that the floodwaters had reached the threshold of the main airstrip on the northwestern end. Houses in the town were evacuated and their occupants struggled to find higher ground.

A decision was made by the authorities to evacuate everyone out by air and the DCA Flight Service Unit duty officer radioed Townsville for immediate assistance. Meanwhile, the Normanton hospital had become completely isolated and the transfer of patients by boat to the airport was far too dangerous in the swirling waters. Fortunately, a helicopter was on hand to run a shuttle service to and from the hospital. An Air Force De Havilland Caribou CCO8 heavy transport aircraft arrived from number 35 Squadron RAAF based in Townsville and began to fly the hospital patients out from the Normanton airstrip. In the early hours of the morning, floodwaters began to pour into the main street, threatening the huge Burns Philp store. Foodstuffs were hastily removed to the airport equipment shed which stood on slightly higher ground and forty men worked continuously to clear the stock, using two trucks until daybreak. Another group of locals transported drums of aviation fuel from the nearby fuel depot to the airfield tarmac. Two RAAF C130 Lockheed Hercules transport were made available from Darwin, but had to be rejected because the Normanton airstrip had now become too soft to support the weight of these huge aircraft. The slightly smaller Fokker Friendship FK27 from Ansett Airlines and TAA were able to fulfil the task with little room for delay or error, which was a credit to the crews who flew them. They were supported by the Bush Pilot DC3s and other smaller fleet aircraft from the same company. An assortment of light aircraft and military helicopters provided the back up needed to hasten the operation as time began to run out. The helicopters provided sweeps of outlying stations by providing supplies for those stranded or by airlifting them out.

The Civil Aviation Regional Flight Service Supervisors in Brisbane requested Normanton Flight Service Officers to remain on duty for a twenty-four hour communications support for as long as they safely could. This overtaxed the already exhausted local officers and arrangements were made to fly in relief officers from the Townsville Flight Service Unit. Local observations indicated that the airport might not become flooded, even though the site was virtually surrounded by still rising waters. The permanent Flight

Service OIC was Harry Telford who had been away at the commencement of the operation on annual recreation leave. On hearing of the situation he joined former OIC Ross Waller and a small group of technical and ground staff personnel who flew in from Townsville as a relief force for the staff already there. By a small miracle, the Normanton Flight Service Unit was able to remain open for the duration and provided weather reports and other messages on behalf of other agencies who no longer had any communications links. When the water finally receded, the township was left with a huge clean up operation and residents began to filter back to repair their damaged homes. Jim Stack had only been relieving at the station to cover Harry Telford's annual leave and finished his tour as a relieving officer of the Normanton Flight Service Unit with some unforgettable memories of his temporary transfer! The dedication of these officers during this outback emergency was typical of Flight Service Officers and should not disappear without recognition.

Weather conditions in Northern Australia can be quite perverse and often caught the unwary. Air searches for missing locals were quite frequent, especially during the wet monsoon season. Mostly, searches were conducted for missing fishermen or leisure boats in the foulest of weather conditions. Sometimes the conditions were so bad, that the volunteer air observers became very ill with airsickness problems. I often used to wonder how anyone could be caught at sea in such appalling weather conditions. Clouds hung down in grey fibrous layers. The water below rose and sank in tremendous troughs and peaks reflecting the same colours as the clouds giving the illusion that we were floating between two shrouds of silk. White caps frothed at the top of each wave and blew in a huge mist in the direction of the wind. Finding survivors or their rubber rafts in these conditions would fall nothing short of a miracle. The gloom of the conditions combined with a forlorn feeling that we are not going to find anyone today and was quite demoralising. We were so far out from the mainland the mission seemed almost a waste of time, except for the fact that we had to be sure that no one was stricken in the area assigned to us to search.

The autopilot would not operate safely in the turbulent conditions and I manually controlled the aircraft hour by hour. Sometimes, when conditions calmed for a few minutes, I re-read the briefing sheet given to me by the Townsville Rescue Coordination Centre prior to our departure.

Townsville Operations and Meteorological Briefing Office 1986. The Operations Flight Planning Counter is on the left with the Meteorological Briefing Counter in the centre of the picture. Author

I checked to verify that I was maintaining the patrol in my assigned area and had not strayed into an area assigned to another aircraft. This was important for two reasons. Firstly, the weather prevented us from seeing each other at the boundary and since we were all compelled by the conditions to fly at 500 feet, the risk of collision was quite real. Secondly, the integrity of the search depended on navigating the aircraft accurately so that the probability of detecting the survivors was not eroded. After all, if you failed to fly an assigned portion of the area there is a possibility that the survivors may be in the portion you missed. There was also little point in flying over an area that had already been covered by another aircraft. The search area was huge and well beyond the capabilities of the resources available and the RCC had no option but to rationalise our efforts in the daylight hours that were available. A visual search at night could not be considered, because the cloud base was too close to the water for search aircraft to continue, without jeopardising the lives of the searching aircrews. Torrential rain began to fall so heavily that I sometimes worried that the integrity of the windscreen would fail. The integrity of the visual search deteriorated and I could no longer fly safely at this level. Reluctantly, we pulled up to a safer altitude in the stormy murk and fervently hoped that we have not missed anyone trying to attract our attention below.

We contacted Flight Service and asked for an appreciation of the weather back in Townsville and the news was not good. We assessed our fuel situation and decided that we had no choice now but to return to base. Off to my left I could just see a twin engine aircraft flying with all its lights on to make it easier to see in these conditions. It carried out a sweeping turn to the left and climbed out of its assigned sector also heading back to Townsville. The violence of the turbulence made flying very unpleasant indeed. Off to my right a bolt of lightning lit up the area with a viciousness that caused my crew to become alarmed and restless. As I was not equipped with an on-board weather radar I hoped that I would not inadvertently bumble into the teeth of one of these storms.

Unlucky aircraft in the past have been structurally damaged and broken up in flight by these conditions with fatal consequences for the occupants. I maintained the designated lowest safe altitude for the area and computed that I had only 45 miles remaining back to Townsville on the direct track. Flight Service must have detected the tiredness in my voice and obtained a clearance from Air Traffic Control without me asking for it. With a headwind of 25 knots, I knew that I would be home in 22 minutes, but would need to fly the electronic instrument approach to land on Runway 19. This was because we would not be able to see the ground and therefore dangerously below the level of the nearest hills if we could not see where we were going! The aircraft bounced violently in the turbulence, which worsened from the mechanical effects of the terrain as we came lower. Air Traffic

Control could see us on the radar and we descended to the minimum altitude allowable on the electronic approach. We could just see the runway lighting ahead and with the final landing checks completed, we rolled safely along the bitumen to the parking area with a collective sigh of relief.

We trudged in the pouring rain back to the Rescue Coordination Centre for a debriefing with the search and rescue officers. We confirmed that we had covered all of the assigned area except the sector where the weather conditions prevented us from continuing safely. This would have to be scanned when the weather permitted us to go out again. Darkness fell and we examined the weather reports for the following day, then tried to get some sleep before our next departure. However, the missing boat was located only ten miles from the coast before we had time to leave and an RAAF Iroquois helicopter picked up the survivors, thereby closing another saga in our search and rescue journals.

In addition to qualifying as a Search Pilot, I also obtained a Search and Rescue Dropmaster qualification. As a result I was often employed in this role for the Department of Aviation in Townsville and was seconded to the National Safety Council of Australia (NSCA). The role of the Dropmaster was to control all aspects of supply dropping to those caught in survival situations both on land and at sea. This included a wide range of activities including boating accidents, lost bush-walkers, aircraft crash situations, police support and defence incidents. The bulk of the work in Townsville was for marine accidents especially in the vicinity of the Great Barrier Reef. On rare occasions we became the victims spending several hours in very bad weather conditions as the result of a hoax prank. One night we were tasked to look for a missing vessel called the "John Sands." Nothing on this mission seemed right and we began to put evidence together as we flew the search patterns. We suddenly realised that John Sands was the registered name of a well known gaming company and that some unbalanced person had decided that this was a good basis to send us out on "a game." Fortunately, the Queensland Police had also become suspicious and with an excellent piece of detective work not only worked out what was actually going on but were also able to locate the offenders and make arrests. Wearily and somewhat disgusted, we returned to base.

Only a few hours later we were off again on a search for a real boat that was sinking in the Coral Sea northeast of Townsville. Light conditions had deteriorated and the integrity of our visual search was significantly reduced because we were so tired from the hoax search only hours before. Unfortunately, we were the only SAR equipped aircraft available at that particular time and we therefore had no choice but to fly the mission. Flight Service kept us up to date with changing operational and weather conditions and we were thankful for the protection that they afforded us while we were operating so far out to sea. On this occasion, we located the stricken vessel and discovered that the crew were already in the water. We turned the aircraft into wind and opened the rear doors ready for the drop. As Dropmaster I prepared the drop sequence with three marine supply containers attached in a chain with a few metres of rope between them so that they did not float off individually. A life raft was attached to each end of the chain and connected so that when I jettisoned the package, a life raft was ejected from the aircraft first and inflated as it fell. The first marine container was attached to this raft and jettisoned quickly in sequence with the next two containers. Another life raft was attached to the last marine supply container and also inflated as it fell from the aircraft. The two life rafts and three supply containers then fell in a row into the water. The life rafts were blown along in the wind dragging the three containers between them. The idea being that if they were dropped properly upwind of the survivors, they would drift in the wind towards them. The assembly formed a horseshoe shape on the water so that the survivors only had to wait for the rope to pass by and grab it to haul themselves along to the rafts. When safely in the rafts, they could pull the supply containers and open them to obtain medical supplies, food, water and survival equipment including radio transmitters and emergency locator beacons. Once settled, the survivors were at the mercy of the wind and sea currents. Their hopes for rescue were now pinned on the operation of the electronic survival beacon and that an aircraft would not only hear it, but have the fuel time to actually conduct a homing on its radio signal.

Author carrying out the duties of Dropmaster during an actual air search in the Coral Sea North Queensland 1985 aboard an NSCA Nomad aircraft. Photo Author

In addition to the life raft and marine sea container combinations, the Dropmaster could also use a device developed by the United States Coastguard known as a "Helibox." This comprised a simple box made of strong cardboard with the flaps at the top folded down at an angle on each of the four sides so that they provided a helicopter blade arrangement. The blades were secured with string to prevent them from blowing upwards as the box was falling from an aircraft. This was a very clever design that allowed a box complete with survival items loaded into it, to be thrown from an aircraft and literally spin towards the ground like a helicopter blade operation. The spinning effect arrested the rate of descent, such that even without a parachute the box could arrive on the ground near survivors without damaging the contents.

On one occasion, I successfully dropped a box near a ridgeline for the survivors of an outback accident. I had placed a portable radio in the box and turned it on with the volume control turned fully up. I had also tuned it onto the frequency that we were using in the aircraft. This was to avoid having the survivors interfere with the radio controls and render it useless through misuse. We banked the aircraft in a very tight turn around the spot where the Helibox and fallen and watched as one of the survivors ran towards it. I timed my transmission to perfection. As he reached the box I started transmitting. "DO NOT TOUCH THE BOX OR THE RADIO INSIDE. IF YOU CAN HEAR ME, FACE THE AIRCRAFT AND WAVE WITH BOTH HANDS." The survivor stopped suddenly then turned and faced the aircraft, waving with both hands. I called again. "AS YOU CAN HEAR, THERE IS A RADIO IN THIS BOX. THERE ARE SOME OTHER SMALL ITEMS THAT MAY BE OF USE TO YOU UNTIL WE CAN GET A GROUND PARTY TO YOU. IN THE MEANTIME TAKE THE RADIO OUT OF THE BOX, BUT BE VERY SURE NOT TO TOUCH ANY OF THE CONTROLS OR WE WILL LOSE CONTACT WITH YOU. IF YOU HAVE RECEIVED THIS MESSAGE WAVE AGAIN WITH BOTH HANDS." The survivor faced the aircraft again and waved with both hands. I smiled. He must have got a shock when the Helibox started talking to him! The rescue cycle goes on and there will always be people who get into trouble. But I am richer for the experience and hopefully, the knowledge gained will be put to good use again if the time comes again.

Letter From New Guinea

Radio communications have been used extensively in Melanesia since 1939. The area encompasses Papua New Guinea (PNG), The Solomon Islands, New Caledonia and Fiji. Originally, PNG was known as New Guinea while Vanuatu was known as the New Hebrides. The systems used paved the way for Aeradio in Australia and I have taken the liberty of adding some technical details in this chapter for important historical reasons.

The physical geography of these areas contains tremendous variations of scenery, featuring impassable mountainous regions down to the swampy coastal plains and numerous islands. Aircraft were the only form of transport able to service these areas quickly. As a result, the aviation industry developed rapidly in PNG providing passenger air services and air transportation for goods and supplies. Aircraft also became extensively used in this region by Australian Government agencies and various international missionary bodies.

Radio networks using High Frequency bands began providing communications support initially for the government agencies, particularly for District and Patrol Officers. Mining companies had already begun extensive explorations of the highlands region in PNG prior to 1939 and discovered gold in the Bulolo River area. Huge Junkers G31 transport aircraft powered by three engines were operated by Guinea Airways in support of mining operations and began transporting dismantled sections of huge river dredges from Salamaua and Lae into the Bulolo Valley. There, the dredges were rebuilt and started operations by processing the gold enriched gravels and silts of the river system. The Junkers aircraft were able to lift approximately three tons of equipment per trip until seven or eight of the dredges had been transported to Bulolo. This was an incredible feat for aviation in this period, considering the primitive aircraft types available. When each dredge was fully reconstructed, they weighed between 200 and 300 tons. Eight such dredges operated in the Bulolo area until war broke out in 1939. Gold mining operations were carried out mainly between Lae and Edie Creek that was located near Wau. Three years later, the Japanese bombed the busy harbour at Rabaul, a commercial centre located on the northern end of the Gazelle Peninsula in New Britain. This action resulted in control of the allied civil aviation industry being transferred to the direct control of the Royal Australian Air Force. At that time, a wide range of very interesting flying machines were available in this region. They included Junkers W34, Ford Tri-Motor, Stinson Reliant, Boeing 40-H-4, DH50A, DH66 Hercules and DH84 Dragon aircraft. Qantas Airways aircraft were requisitioned by the war effort and were employed to begin evacuating allied personnel who were being forced to flee from the invading Japanese forces. The evacuees were flown to safety in Australia via an aerial shuttle service that closely resembled and airborne version of the military sea evacuation seen at Dunkirk in France earlier in 1940. A short while later, Qantas aircraft and crews were also used to help establish a system of coast watching and supporting communication networks around the Melanesian area. These activities directly supported the Australian armed forces intelligence systems that were monitoring all Japanese land sea and air movements. In 1944, the New Guinea Air Warning and Coast-watching network set up an extensive and reliable radio system operating across the entire area, with a main control station operating from Port Moresby. Four basic network stations were established at Nadzab, Dobodura, Milne Bay and Gusap. Information fed to Port Moresby from these stations was relayed to the Australian mainland authorities in North Queensland and Canberra. The operational system of radio networking formed the basis of the Aeradio systems that were also developing in Australia.

The wartime spotter units were equipped with the now famous AWA type 3B Teleradio transceiver. The equipment was considered portable, though not by modern post-war standards and consisted of a transmitter, receiver and a loudspeaker box. This equipment was powered by a 12-volt accumulator with a generator powered by a small petrol engine. The transmitter and receiver cabinets each measured approximately 30-cm square and 60 cm in length. The portable system required a minimum of sixteen men to carry it across the inhospitable region in which the units were operating. Fortunately, before the outbreak of war, the AWA unit was already popular in these remote tropical areas and some one hundred or so were already installed there. They were secured from aircraft owners, mines, trading companies and plantation owners before the Japanese could get to them. The Teleradios were known for the relative simplicity, robust construction and compact size. They were highly versatile and could be operated using continuous wave (CW Morse code) as well as voice modulation using a microphone or handset. These features rendered the unit so popular that it was soon used extensively in the Dutch East Indies, the Solomon Islands and in outback areas of the Australian mainland. Telephone systems in PNG were very poor and radio systems were mostly used instead. When users wanted to contact other centres, they simply called the control station at Port Moresby or Rabaul and obtained a connection in a similar manner to a telephone exchange operation. This service also used AWA equipment and extended to providing a relay function to Sydney or Melbourne or even to centres in the worldwide telegraphic and cable system. Within PNG, radio-telegrams could be sent by booking a line at the nearest Post Office. Thirty of these were operated by the Department of District Services and Native Affairs (DDS & NA). There were four main communications public correspondence links that were operated by professional radio officers, using high speed Morse code circuits similar to the ones that were used by Australian capital city post offices to communicate with suburbs and country townships. These communications links were essential to the rapidly developing commercial centres. The coastal towns of Lae and Salamaua and the inland towns of Bulolo and Wau were expanding rapidly as a result of the gold mining operations. Typically, the township of Lae had actually grown from a mission station that had been established by Lutheran missionaries from Germany.

After WWII, the Placer Development Company from Canada resumed operations in the Bulolo Valley and by 1951 had built an airstrip from crushed rock from the dredging tailings. War had ravaged the area and roads were reconstructed by the Public Works Department in quite perilous conditions on the steep slippery terrain. Works department employees Jack McGrath, grader operator Charlie Whitfield and bulldozer driver Bill Bradley became well known identities as they worked up and down the valley and other road improvements soon followed linking Edie Creek to Mount Kaindi. Mt Kaindi was used to site a repeater station for the Posts and Telegraphs department and this became a vital link in this area. Some locations were quite inaccessible by road, despite modern techniques in road-building and very soon modern commerce became reliant on air operations for faster access. Flying conditions in Papua New Guinea are unique in the world. The steeply rising mountain terrain combined with the hazardous PNG tropical weather conditions made aircraft operations quite perilous. Mountain airstrips were cut into steep terrain that featured dangerous one-way valleys and this required pilots to fly flawless approaches to land on every trip. Failure to do so would have resulted in a high risk of fatal consequences. The mountains were too high for light aircraft to climb over and local pilots had to use all the skill and local knowledge they could muster in order to survive. Flying operations usually commenced at daybreak. This was to maximise flying hours before the afternoon storms, which were a feature of the local weather conditions nearly every day. The storms usually developed at an alarming speed and closed valleys used for aviation access within minutes. Survival was dependent on reading the vital signs that a storm was starting. Many pilots have been killed by the vagaries of the unstable conditions in PNG after finding themselves in narrow valleys, unable to out-climb the mountain peaks and then flying into cloud unable to navigate clear of the rock-walls looming unseen. Inevitably, they have hit one side or the other valley walls and crashed. Air Operations in PNG increased in 1947, when the Australian Department of Civil Aviation (DCA), recommended to the Commonwealth Government that the PNG Provisional Administration required air support in the form of amphibious and transporter operations. This was to take the form of Qantas Catalina flying boat services. The services required connected Port Moresby, Kikori, Lake Kutubu, Daru, Rabaul, Namatanai, Kieta, Green Island, Talasea, Gasmata, Samarai, Misima, Kiriwina, Milne Bay and Woodlark Island. Non-amphibious Douglas DC3 transport aircraft were used to provide services between Port Moresby, Lae, Rabaul, Kavieng, Manus Island, Madang and Wewak. With the commencement of these services, western civilisation came to PNG at an incredible rate. The "Bird of Paradise," passenger services commenced in 1948, flying regular public transport runs between Sydney and PNG using the faithful Douglas DC3 transport aircraft. This aircraft usually carried twenty passengers, two pilots and a cabin attendant. The northbound flight from Sydney usually stopped overnight at Townsville and made other refuelling stops over relatively short hops via Cairns and Cooktown while keeping contact with Aeradio stations along the way. The reason for the frequent refuelling stops was to take advantage of the fuel supplies available and also to ensure that the maximum fuel was on board to allow for any weather diversions at Port Moresby. After landing in PNG, the weary travellers found themselves being flown in tiny DH83 Fox Moth aircraft into very remote localities. This was a daunting prospect for passengers used to city comforts especially since they were obliged to sit in a tri-angular shaped front cabin, while the pilot sat up behind them in an open cockpit, exposed to all weather conditions. At remote locations, the passengers were often coerced into starting the engine by swinging on the propeller, while the pilot sat in the cockpit coaxing the engine into life by manipulation of the controls. A year later in 1948, the rapid development of civil aviation activities in PNG resulted in the establishment of a Flight Information Region (FIR) operating to International Civil Aviation Organisation (ICAO) guidelines. ICAO standardised the procedures used by its members and very few countries refused membership. Now pilots could expect the same signals and procedures wherever they flew in the world with significantly reduced chances of misunderstandings. The purpose of the FIR was to provide an effective administration for the provision of operational information and also search and rescue alerting services for all aircraft. Aeradio stations were established at Port Moresby, Lae, Rabaul, Finschafen and Madang. They were all staffed by professional Radio Officers who provided communications services for aircraft and radio homer responder services to assist pilots with their navigation. Port Moresby became the hub of operations, although Finschafen had earlier been considered because of its superior radio and navigation aid facilities. This equipment had been abandoned by the departing American armed forces after the war. A visiting Australian Department of Civil Aviation technician was amazed at the quality of the equipment which included racks of the famous HRO or AR7 style communications receivers. The equipment was rescued by the Australian authorities who put it to good use in order to provide radio facilities at the new Aeradio stations being formed in the region. Aeradio units were eventually established in primitive circumstances at Bulolo, Wewak, Momote Island, Kavieng and Nauru Island. Nauru was staffed by the Australian Overseas Telecommunications Commission using officers from its coastal radio stations. Control towers were established at Port Moresby and Lae and later at Mount Hagen, Goroka, Madang and Wewak. The first Lae control tower consisted of an old shack that sufficed until a proper tower could be constructed.

Aerodrome control tower and communications centre believed to have been used at Port Moresby Airport PNG. The living quarters and equipment room were on the ground floor. (Civil Aviation Historical Society)

International Radio call signs were officially assigned to each station and promulgated in official circulars to all operators. The responsibility for allocation of radio call signs rested with the International Telegraph Union (ITU). In Australia, aircraft were issued with call signs using five letter combinations of the alphabet while aeronautical ground stations used four letter call signs and international public correspondence circuits used three letter call signs. Amateur stations used either five or six figure and letter combinations and radio beacons used two or three letter call signs on MF and VHF frequencies.

Station	Allocated Radio Call-sign
Port Moresby	VZPY
Lae	VZLA
Rabaul	VZRB
Bulolo	VZBL
Finschafen	VZFN
Wewak	VZWK
Madang	VZMD
Momote	VZMO
Kavieng	VZKV
Nauru Island	VZNI
Townsville	VZTL
Aircraft on initial contact with a ground radio station	VZYZ

Frequency measurements of the PNG aeronautical stations to ensure the accuracy of the transmitters and receivers was performed monthly by the Post Master Generals Office (PMG) frequency measurement centre in Townsville. Tolerances were very accurate and stations identified transmitting on frequencies that were out of limits, were required to effect corrections immediately, sometimes while the measuring station waited on the line. In addition, Port Moresby also used a modified call sign of 'VKY' if it transmitted on frequencies assigned to international networks. International radio communications from point to point used morse code for liaison purposes between Port Moresby, Dutch New Guinea (now known as Biak) and also with Guam. Persistent interference often occurred on aeronautical frequencies and was reported to the Head Office of the Department of Civil Aviation in Melbourne who relayed the complaint to the offending country for consideration and immediate corrective action.

When aircraft called a ground Aeradio station they were not always guaranteed an answer. This was because HF radio conditions suffered from a wide range of interferences especially in tropical areas. Aeradio stations listened carefully to see if the station called by the aircraft was answering. If a reply was not heard within twenty seconds then the station that could hear the aircraft would answer the call and then relay the information to the required Aeradio station by alternative means. HF radio conditions could present some difficulties at times and stations relayed messages to and from aircraft until radio conditions became normal again. An aircraft taxying in Rabaul for example may well have been able to contact Darwin on a particular HF frequency but not able to hear anyone else. In this example, Darwin would answer the aircraft and then relay the information to the correct station by using direct voice liaison or by sending a message on the teleprinter system. One of the Aeradio officers working in Port Moresby was Mr C. B. (Syd) Hulse who recalled that Port Moresby often communicated directly with Qantas DC4 airliners flying over Japan on their way back to Australia. While I was operating as an Aeradio officer in Darwin on the South Eastern Asian HF network 3 (SEA3), I regularly provided a relay for Qantas Boeing 707 jetliners taxying for departure on the ground at Singapore! They were often frustrated, because they could not contact Singaporean or Indonesian Aeradio stations and would therefore call Darwin on a higher frequency to ensure that they were in radio contact with someone in case any emergency occurred. In the early morning at Darwin, we would be amused by stations all over Asia and the Middle East operating on mutual frequencies, calling like amateur hams, to see how many distant stations they could contact before the ionospheric conditions settled down for the day after sunrise. Despite these early morning distractions, the PNG Aeradio stations provided a very good traffic and operational information service to aircraft and an effective search and rescue alerting service for the Port Moresby Flight Information Region. The regional boundaries encompassed an area between 14 degrees south and 03.30 degrees north and a western boundary aligned with the Indonesian international border and a series of complicated coordinates to the east over the Pacific Ocean. Some Australian flights were subject to special procedures because the destination radio facilities were not suitably equipped in accordance with the required ICAO standard. Qantas pioneered a DC3 passenger service between Honiara and Nauru but Port Moresby Aeradio was used on a regular basis to maintain a direct communications contact with the aircraft and provide the required operational and traffic information services. They were also expected to provide search and rescue alerting if the aircraft failed to report at the mandatory positions. The aircraft crew consisted of two pilots one of whom was also required to be the holder of a first class navigator's licence. The PNG Aeradio supervisor often flew with them and performed radio officer duties using morse code on international frequencies to give position reports and also report safe arrival. During the early 1950s the RAAF were still responsible for performing air searches for missing aircraft and boats. Any Aeradio unit first becoming aware that an aircraft was missing, immediately

230

notified the RAAF authorities in Townsville. During this period the RAAF used the GAF (Avro) Lincoln bomber to patrol the search area and if necessary drop supplies to survivors. Unfortunately, these aircraft invariably became unserviceable in flight and were compelled to make a premature return to base. The responsibility for search and rescue was later transferred to the Department of Civil Aviation who were considered better placed to find civil aircraft resources for search and rescue purposes. Civil aircraft were normally available in larger numbers to perform the required searches for missing aviators and seamen. Department of Civil Aviation Air Traffic Control Centre Operational Control Units became the controlling authority for such searches. As commercial, missionary and government operations increased in PNG, the aircraft numbers intensified and Air Traffic Control services were increased at Port Moresby, Lae, Madang, Goroka and Mount Hagen to control the traffic problems that occurred, especially during daylight hours. Aeradio officers increasingly found that their duties included liaison with a large number of other services, simply because most Aeradio stations were available 24 hours a day and could be relied upon to provide assistance to anyone in trouble, whatever the circumstances.

On Sunday 21 January 1951, the local churchmen in Popondetta must have thought that the end of the world had come when the countryside erupted in a volcanic explosion that was so sudden and violent that 2942 people lost their lives. At the centre of the eruption was Mount Lamington situated 20 kilometres south of Popondetta township. One side of the volcano, which stood 5509 feet above sea level, had been completely blown away and the landscape for miles around was literally flattened. Jeeps and other vehicles were reported being tossed in the air like little toys and the jungle disappeared in a coating of grey pumice and debris. The area suffered almost total destruction and the disaster was of such a magnitude that Australia was compelled to provide immediate assistance and initiate emergency relief programmes. Paramount to the relief exercise, was the need for continuous radio communications and volunteer radio-men were called to service. DCA Aeradio officers were specifically requested to assist, especially those with any experience in PNG and also preferably, those with any experience in handling public correspondence channels. Mr C.B.(Syd) Hulse was one of the volunteers. Syd had just returned to Australia after completing a tour of service in PNG and had only been back for twelve hours when he was alerted to assist with the relief radio operations programme. He had been a telegraphist at the Post Master Generals office in Melbourne prior to transferring to the Department of Civil Aviation. Syd had several radio qualifications which were more than suitable for the relief operations and so he boarded the next aircraft available and returned to Port Moresby's Jackson Field airfield to commence a six week contract. He finished up staying for six months before returning to Essendon airport in Melbourne. However, the stay in Australia was again short lived, because after only four weeks he was required back in New Guinea. This time he remained for nearly ten years and left in June 1960 on another transfer, this time to Brisbane Aeradio at Eagle Farm airport. He recalled some of the events he experienced during this period:

"After the volcano had erupted, I saw the area from the cockpit of a Gibbes Sepik Airways Norseman freighter aircraft. I could hardly believe the devastation below. I was seconded to the PNG administration Post and Telegraphs communications centre, to help with the sudden increase in radio telegraphic traffic between regional centres and the two emergency field units that had been established at Popondetta and Oro Bay by the Department of Civil Aviation. These two units were manned by Aeradio officers at each location. They were required to provide radio communications with both DCA and the Administrations communications centre to report on any further seismic shocks or eruptions. When DCA Port Moresby closed down at 9pm each day, its responsibilities were taken over by the Administration's communications centre. I was rostered for permanent night shifts for a number of weeks on alert. The two officers at Popondetta and Oro Bay lived and worked in the most appalling conditions

Doug Whitfield

for several weeks, living not unlike the Australian Army soldiers in the wartime New Guinea campaigns. When the two officers were finally relieved, they returned in very poor health because of an unsatisfactory diet and also because they had suffered badly from malaria, dengue fever and tropical ulcers. Never the less, they had remained on duty without any fuss and with a devotion to the job that was well above the call of duty. It is an irony, that these men did not receive any official recognition for their work, although the heads of certain Government Departments did receive awards.

Aeradio officers generally operated with a strong sense of team spirit. They enjoyed life to its limit and worked hard in very harsh conditions by today's quite soft standards. Staff movements were more flexible then than the union controlled idiosyncrasies of today. An officer might only be given a few hours notice to board an aircraft and go to another station as a temporary relief measure for staffing. Some transfers may only have been a week or so, but others certainly had to stick it out for six or seven month's duration. It was all part of the job and living conditions varied greatly from place to place. In Port Moresby we lived in the RAAF Officers mess and living conditions were very good and quite different from Lae, Rabaul or Bulolo! The accommodation at Lae was an old wartime building built on the edge of the airstrip and presented a noisy and spartan environment indeed. Rabaul was similar, except that the DCA accommodation was located in the town away from the Aeradio shack and the airstrip. At Madang, the walls of the accommodation was constructed from woven grass on wooden frames with a tin roof above. After a night shift from 3am to 10am, sleep was almost impossible while DC3, DC4 and JU52 transport aircraft taxied up and down revving their engines. Eventually, in the late 1960s, new housing was built and modern radio consoles replaced the old shack equipment. Aeradio officers who had been at Lae would remember that the wet conditions were so bad at times, they had to have rubber boots to keep their feet dry and also to insulate themselves against an electric shock while handling the radio equipment!"

Duties for Aeradio officers located in Port Moresby included a requirement to staff a radio and reporting outstation on Daugo Island located approximately 20km southwest of Port Moresby airport. The Island was used as the alternate airfield to be used for diversion if the weather at Port Moresby precluded a safe landing by Qantas airliners flying on the 'Bird of Paradise' service which normally arrived at 6am. The duty Aeradio officer was required to obtain food rations from the mess kitchen and embark from the Fairfax Harbour marine base aboard the DCA crash-launch at 4pm on the day preceding the expected arrival time of the aircraft. This was because the trip to Daugo Island took about 45 minutes from the mainland coast, which was a distance by sea of only 10km. However, there was a lot to do on arrival. The Aeradio officer waded ashore and set up camp in a corrugated iron shed, which housed the Aeradio shack and a combined bedroom/kitchen. He then started the diesel driven power supply and tested both the Non-Directional Radio Beacon (NDB) and the Collins transceiver. If this was satisfactory he rested and commenced duty again at 2am. Radio contact was established with the Port Moresby control tower on a frequency of 3023.5 kHz and weather reports were to be sent and received every 30 minutes. Strangely, Qantas policy precluded any landings at Daugo Island because there were no ground facilities for passenger disembarkation! A ruling was received from DCA Head Office that if the Port Moresby weather precluded a safe approach and landing, the aircraft was to return to Cairns. Daugo Island continued to be staffed on the off chance that an emergency may occur which could place the passengers in undue peril if the option to go there was removed. The DC4 aircraft used on the 'Bird of Paradise' service was later replaced by the Super Constellation L1049 aircraft. The superior performance of the L1049 prevented it from using the runway at Daugo Island because the runway length was simply too short. The use of Aeradio officers therefore came to an abrupt end, which was unfortunate, because the trip to and from Daugo for them, passed by very picturesque scenery. Very often after returning from a shift on the island the Aeradio officer was also at liberty to spend the rest of the day fishing while off duty. The methodology of the fishing activities raised the eyebrows of newcomers as the locals threw explosives off the back of the launch in certain places and then speedily scooped up as many stunned fish off the water's surface as they could before the arrival of sharks. However, the prospect of sharks did not deter local native boys who seemed unconcerned as they swam retrieving as many fish as they could. The Australian personnel needlessly to say, viewed this activity rather warily. Syd Hulse recalled some of the more enjoyable experiences of the time.

" *Sometimes, when I had some spare time, I would go for a ride with the coxswain of the duty crash launch when it was checking the water landing areas for the departing and arriving flying boats. The launch was powered by twin diesel engines, which gave it a top speed of about 20 knots. It used to take up its station on the starboard side of the flying boat as it accelerated on its take off run. The aircraft did not take long to outrun the boat and gracefully swept away on its departure. A variety of amphibious aircraft operated from Port Moresby, including the large Sandringham flying boats, which used to fly a regular public transport service from Rose Bay in Sydney via Brisbane and Cairns. I flew in them several times and they were a delightful aircraft, fitted inside with cabins, tables and chairs with each cabin able to seat four persons. It had big windows that gave an excellent view of the area over which you were flying. At Brisbane we landed on the Brisbane River at Hamilton, where crash launches were busy ensuring that there were no obstructions in the water. Usually the flight flew below 10,000feet along the coastline on the southbound trip and stopped at Cairns for breakfast while the aircraft was refuelled. The flight crew always dressed in naval style uniforms and pilots with extra qualifications such as a Navigator's ticket wore a gold bullion star above their rank stripes. Some of the older captains also had commercial radio operator's qualifications and the obvious advantage of this expertise was that they could perform each other's jobs on long flights if someone felt like a break from their duties. The individual crews were well known to the Aeradio staff on long flights to Guam and Kure during the Korean war and later when the L188 Lockheed Electra aircraft were acquired for the air route which ran from Manila to Port Moresby."*

The biggest commercial operation in PNG was undoubtedly gold mining. After WWII, mining restarted with a formidable momentum. One of the biggest companies to enter the arena was the Placer Development Company of Canada, which recommenced the Bulolo Valley operation still using the huge dredges to mine the gold from the river areas. A large workforce was employed and the company constructed an excellent all weather airstrip from the dredge tailings. In 1951 Syd Hulse was posted to Bulolo and began operations in the Aeradio unit established there.

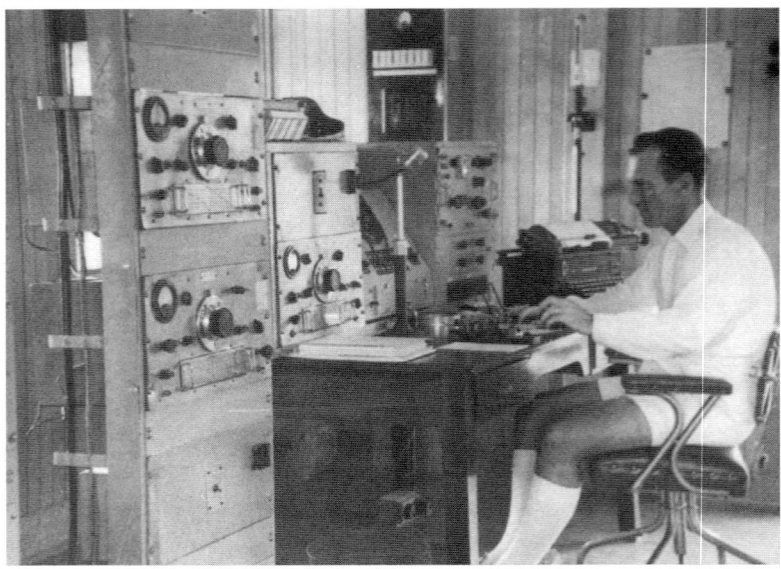

Mr C.B. (Syd) Hulse operating the Bulolo PNG Aeradio station in the early 1950's. Note the AR7 HF receivers on the rack left of the table. A mercury barometer can be seen on the wall in the background. (C.B. Hulse)

The Placer company constructed a building, which roughly resembled a control tower, and it was elevated near the airstrip to afford a good view of the aircraft movements. Qantas operated daily local flights and flew between Bulolo and Lae using DC3 transport aircraft and the lighter DHA3. Large amounts of gold were airlifted out on the aircraft and taken to the Bank of New South Wales in Lae. In a relatively short time, Placer had managed to extract over thirty million dollars worth of gold from the valley. The landscape soon began to resemble a lunar surface with craters and cuttings everywhere. Living conditions were harsh for everybody and Syd spent most of his time working and sleeping in the Aeradio shack. He was one of the very few 'non-company,' employees in the valley. The company looked after him however, and provided free meals and staff accommodation if required for any of the Aeradio officers that worked at Bulolo. Syd was forced to sleep at the aerodrome because of the irregular duty hours. In addition to the mining venture, the Placer Company also ran a large plywood factory because they had managed to secure a contract to fell and process the Klinki pine trees from the vast mountainous forests. The work force consisted of approximately two hundred expatriates and approximately five hundred PNG nationals. Communications links between communities were difficult and equipment was often unavailable. However, an unusual system was used to communicate between Bulolo and Wau, a distance of only 17 kilometres over steep mountain ridges and deep valleys. A battery operated portable 75 mHz transceiver located at Bulolo employed a modern Yagi antenna array on a mast, that was positioned, so that the signals could be radiated and physically bounced off the mountainside! In this way the messages transmitted could be received by the groundsman up the valley at the Wau airstrip. The system was reasonably successful and quite ingenious and important weather information could be provided for pilots inbound to the airstrip from locations further down the valley. Wau airstrip is built on a hillside, which is so steep, that aircraft landing there, have to touch down and then immediately apply full power in order to reach the top. Failure to do so, would leave the aircraft stranded at the lower part of the strip and in need of considerable labour to reposition it up at the top again. Once there, a sharp right turn was required to manoeuvre onto a tiny flat area, which was only just suitable for parking. When the aircraft was ready to leave again the takeoff technique required rapid application of full power to get airborne quickly while hurtling down the hillside from the force of gravity. The airstrip location was famous in the 1930s as a base for supplying the local goldfields at Edi Creek. The allies fought the Japanese there in heavy conflicts in WWII and AIF infantry were flown in and out while the front line of the battle was literally moving on the outskirts of the township. After the war finished, the facilities had to be rebuilt at both Wau and Bulolo to enable resumption of commercial activities. The Wau location is very picturesque with panoramic mountain views and much cooler temperatures than those in the provincial city of Lae down on the coast.

The Aeradio unit at Bulolo was equipped with racks of Australian-made Kingsley AR7 radio receivers that were very similar to the American ones left in the Finschafen area. Bulolo used two AMT150 transmitters capable of using voice-modulated transmissions or continuous wave (CW) transmissions using the Morse code. The Kingsley AR7 was a descendant of the legendary 'HRO' receiver, which had been designed in the 1920s by the National Company of Malden Massachusetts in the USA. There seems to be some conjecture about how the name 'HRO' came into being. One source suggests that when the original contract was signed in the United States, the demand exceeded supply to such an extent that the manufacturers were compelled to work at an extraordinary rate to meet the deadlines. The radio receiver became universally known on the shop floors as the 'HOR' for 'hell of a rush!' The name stuck to the model and was adopted and modified to HRO for aesthetic purposes in commercial advertisements.

At the time of development, the US Federal Aviation Authority (FAA) had just been formed and was very busy building a communications network. They selected the National SW5 model radio receiver for use by its ground to air communications stations. Radio technology was developing at a phenomenal rate and the National Radio manufacturing company quickly entered the market with a superheterodyne radio equipped with a radio frequency (RF) circuit, an automatic gain control (AGC), and a crystal intermediate frequency (IF) filter. The technical explanation of these features is beyond the scope of this book, but these circuits were considered to be very advanced for their time.

Close up view of the AR7 HF receiver used extensively by Aeradio stations in PNG. The tuning coils are located behind the calibration graphs and could be removed to change frequency by pulling the two small handles on either side. (Author)

The AR7 and similar style radio receivers were made by a variety of manufacturers. The units employed a plug-in coil for each band and covered an overall frequency range of 1500 kHz to 20 mHz. The coil assemblies were plugged into the front panel and this required the tuning capacitor to be mounted longitudinally so that it was parallel to the front panel. The mounting for this mechanism required a right-angled drive that used the now famous 'PW' drive. The drive was easily recognisable on all radio sets employing this feature, because the large main tuning knob was mounted on the tuning dial in the centre of the front panel of the set. As the tuning dial was rotated, the calibrations appeared in slots in the face of the tuning dial. There was no backlash in the gearing and this allowed accurate calibration and mechanical band spreading. The 'HRO' was very popular and copied by a number of radio manufacturing companies. Kingsley Radio in Melbourne was probably the most famous of these, with its production of the AR7 receiver while Amalgamated Wireless Australia (AWA) produced a similar unit known as the 'AMR-100.' Howard Kingsley Love was born in Australia in 1895 and obtained an experimental wireless licence before he joined the AIF in WWI. He was in the AIF a short while and then graduated as a fighter pilot in the Royal Flying Corps. In the early 1930's Love founded his own small radio company which he called after his middle name. When war broke out again in 1939, Love became involved in Defence work and was able to use his military training to full advantage. He won a contract to design and supply high performance radio receivers for the RAAF and the first of these was the Kingsley KCR-11 which was based on the American National companies model 'HRO' receiver. The KCR-11 was designated the 'AR7' by the RAAF and the first order for approximately twenty receivers was followed fairly rapidly for literally thousands more, most of which were to be supplied for the RAAF, Australian Army and also Dutch Navy use. The Kingsley factory was located in Spring Street in Melbourne, but the premises soon became too small for the level of production required. This was remedied by moving to new premises in St Kilda Road, which was close to the RAAF Headquarters and the Victoria Army Barracks. This move no doubt substantially improved liaison between Kingsley radio and his military customers! The company became involved with the development of wartime radar equipment and other projects with the AR7 receiver that demonstrated a need for better quality ferromagnetic cores and IF transformers. By ill fortune Howard Kingsley Love died suddenly in 1948 and the company was, for some poorly conceived reason, liquidated in the same year.

Harry Gentle was a well-known DCA Flight Service

Supervisor working in Melbourne during the 1970s and was one of the early users of the Kingsley AR7 HF receiver during his earlier days in the Australian Army. The Army used the AR7 as a portable radio, although its portability was confined to being mounted in a light truck, simply because it could not be manhandled, especially in steep and heavily forested terrain. The system components were transportable and included a bulky power pack, a collection of coil boxes and an antennae unit, in addition to the receiver module with its eight glass electronic valves. The receiver required a great deal of constant maintenance, but was none-the-less highly regarded by most radio operators. During the post-war period the RAAF and Australian Army made extensive use of the AR7, which was considered to be ahead of its time, despite the introduction onto the radio market, of the American-built AR88 which was considered by some to be superior in performance. More Aeradio stations began appearing around Australasia and they too made extensive use of the AR7 equipment particularly for the reception of morse encoded messages. Operators like Harry Gentle found that the frequency stability varied during continued operation of the set and required great attention to ensure that the receivers remained tuned to the particular frequencies in use. The tuning procedure was usually performed with the assistance of a Bendix frequency meter but frequency drift problems still continued until the Civil Aviation technicians invented coil boxes with crystal locked frequencies which became permanently employed. Unfortunately, the coil boxes were very fragile and were often damaged from over-enthusiastic handling by operators who were desperately trying to maintain contact with a particular station. If the frequency required a coil box changeover, the rapidity of the box change was often not the most gentle in the heat of the moment. The AR7 dial was not calibrated for specific frequencies and frequency resolution was completed by using a graph mounted on the front of the coil box on the front panel of the receiver. Tuning procedures were laborious and operators often marked the frequencies used on the dial itself. One of the Aeradio duties was to monitor the morse identification from Non Directional Beacons (NDBs) within its area of responsibility. Depending on their location, beacons could be quite difficult to monitor and operators marked the position of the particular beacon on the AR7 dial. The numerous scratch marks made on the dial to indicate particular frequency locations resulted in the dial looking rather scruffy. These limitations were common in all radio receivers during this period and eventually, the faithful AR7 was replaced by the Australian built R20 receiver.

Liaison between Aeradio units was effected by point to point radio contact. This required relaying messages through intermediate stations so that higher frequencies did not have to be used for longer distances. In this way, frequency usage was optimised to avoid traffic jams. If Perth wanted to relay a message to Wyndham, they asked Meekatharra to relay it to Port Hedland or Derby who in turn relayed the message to Wyndham. Similarly, if Port Moresby wanted to relay a message to Brisbane, it was relayed through Cairns or Townsville and Rockhampton or Bundaberg, depending on how ionospheric conditions affected the HF skip distance and how the signal was being reflected back to the earth's surface. The AMT150 transmitters were usually used with a 150-watt output, which proved adequate for normal operations. Liaison with other units was normally conducted on 3mHz and 9mHz depending on the diurnal variations in the ionosphere. Radio reception at Bulolo was normally excellent and the operators in the unit there were able to provide a good service by relaying messages for other units especially to Dutch aircraft operating between Hollandia (now Djaya Pura) and Merauke. The position reports from these aircraft were normally transmitted on 6540 kHz using morse code.

Additionally, a code known as the "International 'Q' code" was used in order to minimise otherwise lengthy transmission times. The 'Q' code basically consisted of hundreds of three-letter groups beginning with the letter Q, and these represented certain messages.

Some examples of this code are shown below

QRM	Radio reception is subject to interference
QNH	Altimeter setting using sea level as datum
QFE	Altimeter setting using the aerodrome reference point as the datum
QDR	Downwind leg or Reciprocal Heading
QDM	Given Heading to fly

As voice modulated equipment began to replace the old morse-code circuits, with plain language transmissions, the 'Q' code became obsolete. However, a few of the codes were actually retained for convenience to minimise transmission times on the voice circuits. Thus a Flight Service or Air Traffic Control Unit would simply say "QNH 1010," meaning, "the standard pressure setting for today to be set on the sub-scale of your altimeter is 1010 millibars."

When I was stationed at the Darwin Flight Service station, the unit was compelled to retain a morse circuit, because our international neighbours to the north had not yet updated their equipment. Weather forecasts, flight plans and coordination messages regarding aircraft in flight, were sent by the morse operator to the island in Indonesia as morse encoded messages. This was necessary because organisations like the Phillips Oil Company operated a Catalina flying boat from Darwin across to Biak, Timika and Ambon. The Catalina was later replaced by a Fairchild FA24 turboprop aircraft that was basically a Fokker Friendship FK27 in disguise! Occasionally, the South Australia and Territory Aerial Services (SAATAS) operated from Darwin across to Dili on the island of Timor using a variety of aircraft from Cessna 185 single engine tailwheel aircraft through to the Beechcraft Queenair BE65 twin engine aircraft. The Department of Civil Aviation provided the services normally applied within Australian territory to these aircraft because they were flown mostly by Australian crews. The aircraft were required to communicate with Makassar Aeradio using a frequency of 8820 kHz, but Darwin was not equipped with a transmitter for this purpose and could therefore only monitor the frequency by tuning and using the AWA CR6-B HF receivers. The morse operator on duty in Darwin sent the routine aircraft movement messages and weather information ahead to the radio operators located around the area, who were responsible for aircraft monitoring in airspace controlled by Indonesia.

I mentioned earlier in this book, that the morse operators had their own 'fingerprint' and were able to identify each other. The same identification methods were used to detect certain Indonesian morse operators. The operators in Darwin were able to identify certain characteristics in the morse messages received from most operators. Some were branded as 'not so good' and others in much better terms. I was idly monitoring the Darwin operator one morning, when the Indonesian operator sent what was known as an interruption code to stop Darwin from transmitting. He then chastised Darwin, by sending a message, which read, "Darwin, please you send proper!" I was mildly amused as the Darwin officer unabashed, muttered 'cheeky bugger,' and instantly replied with, "Roger - please you RECEIVE proper." Normal operations resumed instantly.

Aeradio officers were ambassadors for Australia in their own right and were conscious of the diplomatic obligations while they communicated with many of our international neighbours. They were also obliged to perform a wide range of duties that included, acting as customs and immigration officers when international aircraft were forced to divert from their normal flight plans to unplanned destinations because of deteriorating weather or an aircraft emergency. They were also required to inspect airfields for unserviceabilities and take charge of, or assist with local civil emergencies. Officers were constantly being posted between stations and were usually accompanied by their families. They were able to develop a broad expertise in aviation operations after only a short period of time.

Syd Hulse spent a number of years journeying around Melanesia, notably Papua New Guinea and operated in many interesting and diverse locations. The following is an extract from his verbal recollections of his travels as an Aeradio officer and later as a Flight Service Officer when the service was renamed.

"We worked and lived in some varying standards of accommodation. The mess and sleeping quarters at Madang were fairly primitive and made of grass matting and timber. Both were located on the edge of the airstrip and the noise from aircraft movements made sleeping while off duty day or night almost impossible. Morning briefings for aircrews commenced at 5am and this meant that weather reports, operational information and equipment had to be ready at that time. Idle moments were rare because you were transmitting and receiving messages from aircraft overflying your area of responsibility as well as from the ones departing and arriving at your airfield. Most pilots operating into the PNG highlands tried to complete their flying operations by early afternoon to avoid the tropical storms that were a feature of most afternoons. While I was at Madang, the river levee bank collapsed one night and flooded the entire airfield, inundating the hangars and operational buildings. The airfield groundsman had been working on the other side of the airfield for a couple of days and we were amused to see him swimming across the airstrip to get home the following morning. The airfield was severely damaged and needed to be closed for approximately six months so that it could be rebuilt. Sometimes, very small aircraft with very light loads were able to land during the rebuilding phase, but conditions would not support the larger aircraft operations. Commercial chaos ensued during this period because traders relied almost entirely on aircraft for the success of their businesses.

In the highlands, Goroka was becoming important as a commercial and administrative centre and soon enough an Aeradio station was established at the airport to cater for the increasing aircraft movements there. It was commissioned by the late Mr Maurice Tie, an Australian-born Chinese officer. He was well known in the area and a popular personality who became a commercial pilot in Canada some years later. Tragically he was lost in a fatal air accident there. However, he began operations at Goroka airfield as an Aeradio officer within five days of arriving there, operating initially from a tent. He later organised the building of a typical native hut made from grass complete with a trampled earthen floor. The power supply for the radio equipment came from a small Briggs and Stratton generator. Maurice and his successor Ted O'Meara were pioneers in every sense of the word. They worked long hours, labouring tirelessly and providing services to a rapidly expanding aviation industry in the PNG highlands. These men were typical of the many that found themselves in similar circumstances and deserve honoured places in the civil aviation history of Papua New Guinea. I had an interesting time at Rabaul because the area is subject to volcanic activity and also experiences earth tremors known locally as Gurias, both of which were almost a daily feature. In the 1950s while I was there the Aeradio shack was constructed from timber and tin sheeting. The transmitters were located on the other side of the airstrip and the environment was very primitive indeed.

The Gurias caused the antennae masts to whip backwards and forwards quite violently and water cascaded from the water storage tanks just outside the building. On one occasion, the banks of radio receivers leaned over and threatened to fall on top of me while I was sitting there performing my duties. I was forced to ask the Port Moresby centre to 'standby' while I propped everything up temporarily in an endeavour to stay 'on the air.' Fortunately, the situation calmed down and I was able to continue operations uninterrupted. The Japanese war effort was still very much in evidence around the area. The harbour was still full of sunken Japanese ships that had been bombed by the RAAF during the war. Wrecked Japanese aircraft literally littered the area, especially at the airstrip and salvage companies soon cleaned the area and made large profits from the removal of war surplus found there. The Japanese had constructed a vast cave system into the sides of the hills and used them as military storage areas. The caves were so secure that they survived repeated bombing attacks from allied aircraft.

Several Australian DCA officers were sent on to Finschafen and Kavieng to establish Aeradio units at those locations. I was later sent to Honiara in the Solomon Islands to begin a six-month posting as the Australian Government communications representative and worked basically as a one-man unit. The island was administered by the British Colonial office in London and everything was affected by British attitudes. Honiara is situated on Guadalcanal Island which is very mountainous. There are

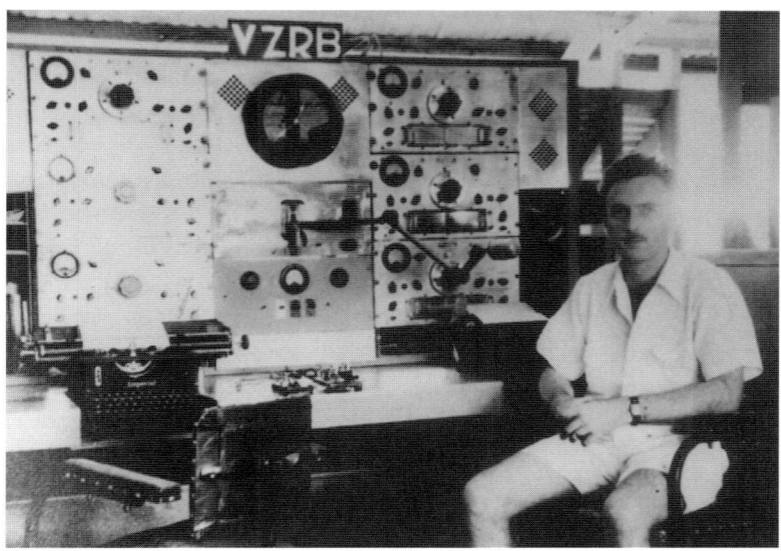

Mr C.B (Syd) Hulse operating the Aeradio console at Rabaul in 1952. The station callsign VSRB is clearly visible. Banks of Kingsley AR7 HF receivers are the main feature of the console (CB Hulse)

terraces rising from the shallow strip of flat lands and this area was the scene of intense fighting by allied infantry and Marine Divisions against the Japanese military forces who drove their troop carriers at full speed onto the coral here to permit rapid disembarkation. The Honiara coastal radio station was a large building set on a terrace some two or three hundred feet above sea level. It commanded an incredible view of the area that included Savo Island where HMAS Canberra had to be scuttled after a vicious night battle during the war. The Royal Australian

247

Navy (RAN) used to conduct memorial services annually on the site. The officer in charge of the Honiara radio station was a New Zealander who had also seen service in PNG as an Aeradio officer. Several Solomon Islanders staffed an extensive island network using voice and morse encoded transmission methods, mostly for weather reporting and administrative business. The station callsign was 'VQJ' and its primary duty was to communicate with ships at sea and to liaise with the Overseas Telecommunications Corporation (OTC) station in Sydney, which used the callsign 'VIS.' OTC is now known as 'TELSTRA.' My specific duties included transmitting lengthy synoptic weather reports several times a day to Port Moresby using a morse circuit and then receiving weather reports from Tarawa in the Gilbert and Ellice Islands. I also had to provide the communications and aeronautical services for the weekly Douglas DC3 aircraft service from Rabaul. We all shared the duties when sickness from malaria occurred and this was a wretched business for those inflicted with the disease. Some of us helped to operate the commercial broadcast station at night and often ran jazz musical segments on Friday evenings, which was a real pleasure because the station was the proud owner of a first class library of music. I soon became good friends with the visiting Qantas aircrews and also with the radio officer from the 'MV Malaita', a trading vessel from the Burns Philp company which operated regularly from Sydney to Honiara via New Caledonia and the New Hebrides before proceeding on to PNG ports. Our main transport on the island consisted of a 250cc motorcycle on which we used to explore a multitude of disused roads and tracks around the island. One of my most vivid memories was riding on the pillion at 70mph through high kunai grass right down the middle of the airport at Henderson Airfield. My memories of my tours of duty in the tropics are very pleasant, especially since I met my late wife Paula in Port Moresby while I was the Flight Service Centre Supervisor there. Port Moresby airport had been named Jackson Field in memory of Squadron Leader John Jackson who was killed as Commanding Officer of number 75 squadron RAAF, while flying Curtis Kittyhawk fighter aircraft against the Japanese in WWII. I finally left PNG in June 1960 and was sent back to Brisbane to work as a shift supervisor in the Flight Service Centre. However, air communications were conducted from Archerfield aerodrome and the message-switching centre was located at Eagle Farm which is now Brisbane's airport. The Supervisors had to alternate duty between each location every second week until all communications facilities were combined at Eagle Farm. Typically, two years later I was posted with my wife and family to Cloncurry where I took over as Officer in Charge. The staff consisted of nine Flight Service Officers and sixteen other staff comprising technicians, electricians and groundsmen. Cloncurry was a very busy station because it provided communications services to all domestic and international flights transiting the area as well as the arrivals and departures of light aircraft operating throughout the Cloncurry Flight Information Area. A close liaison was maintained with the RFDS who operated the De-Havilland three engined Drover DHA3 aircraft on a permanent lease from Trans Australian Airlines. The Cloncurry RFDS base was one of the original bases to commence operations in Australia and when it routinely closed operations for the day, the Flight Service Unit monitored the RFDS frequencies for them in order to provide a twenty-four hour coverage for emergencies. The Cloncurry aviation scene became so busy that a DCA Rescue Coordination Sub-Centre was established to provide support for the Townsville Rescue Coordination Centre that had been responsible for Search and Rescue operations in the entire North Queensland area. It is interesting to note that the Visual Flight Guide (VFG) issued to pilots by the Department of Civil Aviation during this period, included a note in the flight planning section, which stated that:

"Should you consider using the radio network facilities of the RFDS as a means of reporting arrival, or for any other purpose, you must remember that base stations of this network do not maintain a continuous twenty-four hour listening watch. Generally 2-hourly schedules are maintained between 0800 and

1700 LMT Monday to Friday, with a later commencing time of 0900 on Saturday and Sunday. These schedule times vary from base to base and stations may close between 1200 and 1300."

Flight Service was constantly obliged to provide the necessary support for the RFDS during the time their own bases were not manned.

Syd Hulse returned to Melbourne in the winter of 1965 and became Officer in Charge of the Flight Service and Aeronautical Telecommunications School that was located in the Central Training College of the Department of Civil Aviation. He remained there voluntarily until his retirement on 30 June 1989. The Prime Minister at that time was Mr R L Hawke who wrote a letter congratulating Syd for 46 years of dedicated service. During his administration at the College, Syd saw 1400 Flight Service Officers, 500 Aeronautical Telecommunications Officers and 100 Communications Officers graduate. (The author graduated from number 11 FSO course in 1970). Additionally, Syd supervised specialist-training courses for the Nepalese Senior Supervisors under the Colombo Plan and also conducted annual courses for the Australian Antarctic "Expeditioner" radio officers. In all, sixty-six Flight Service courses had been completed during this historic period of civil aviation in Australasia. Syd was invited to participate in visits to Japan, Great Britain, Luxembourg and Hong Kong to evaluate communications and training methods. His observations were used to compare Australian services and ensure that our standards were equal to or better than International standards. The motto of the Central Training College was "In pursuit of excellence." This goal was certainly achieved, though sadly the old work standards have been eroded by automated satellite direct communications and microwave repeater links. With it, the people who gave such dedicated service have passed without notice and this chapter in Australian aviation communications history should never have passed without recognition.

Aerodromes

It would be improper for me to ignore a highly professional body of people who ensure that Aerodrome Standards are maintained in order to maximise safety and where possible standardise all common procedures so that the safety of aircraft using the terminal infrastructure is assured. In the 1950s through to the 1990s most aerodromes in Australia were owned by the Commonwealth Government and operated according to strict licencing standards. Aerodrome Inspectors travelled all around their respective states making sure that visual markers were up to standard and the movement areas were properly serviced and safe for aircraft operations. The duties of the Inspectorate were technically complex and diverse. The qualifications required a very good understanding of aircraft operations, aerodrome lighting, navigation aids, pavement surfaces, obstacle hazards to aircraft such as power lines, tall antennae's, multi-story buildings and a host of matters relating to the Civil Aviation Regulations.

Back in 1921, the specifications and rules for aerodromes were relatively simple. Qantas declared that the specifications for the new enlargement of Charleville Aerodrome in Queensland should be " a minimum area measuring 300 feet by 300 feet with no trees, fences, telegraph or telephone wires around the edges". They further declared that *"if high tree's or the other obstructions mentioned are round the edges, 450 feet by 450 feet is required."* The Controller of Civil Aviation was more conservative and set the minimum dimensions of an aerodrome approved for public use as 350 yards. The required approach clearance for aircraft was set at a gradient of 1 in 10 with an overall slope with a gradient of not more than 1 in 50. The surface needed to be smooth enough to allow a light truck to be driven at a speed 20 miles per hour without any discomfort to the occupants.

A year later in 1922, the Commonwealth Government initially constructed 48 new aerodromes. This had dramatically increased by 1924 to 130 aerodromes and by the end of 1930 this had increased to a total of 62 aerodromes and 122 emergency grounds owned by the Commonwealth Government. 53 of the aerodromes had been issued with the first aerodrome licences. Five years later in 1935, the regular air services operated 12,032 flights and this increased to 47,698 in 1938/39. By 1939, there were now 71 Government aerodromes, 147 emergency landing grounds and 213 licenced aerodromes as well as 7 flying boat bases and 4 alighting areas that had been established between Sydney and Darwin for the Empire Air Mail Scheme.

In 1938, the Department of Civil Aviation (DCA) was created, replacing the then Civil Aviation Board that was formally abolished on 12 January 1939. The Commonwealth Director General of Works, Mr M.W. Mehaffey was appointed as Acting Director-General of Civil Aviation with the initial tasks of transferring responsibilities from the Department of Defence and implementing the responsibilities of the new Department. On 3 April 1939 Mr A.S. Corbett MBE was appointed as the Director-General controlling a staff of 251. Considerable Defence activity began at the outbreak of World War II with numerous tactical airstrips being constructed around the country notably in the northern areas of North Queensland and the Northern Territory. Numerous airstrips were built south of Darwin on the Stuart Highway where large military support camps developed. Despite the horrendous loss of human life, the war was responsible for some incredible technological developments in aviation, especially with the production of sophisticated aircraft using very advanced navigation systems and demanding the same sophistication in airfield facilities.

Aside from the civil aerodrome developments at Sydney, Melbourne, Brisbane, Cairns, Townsville, Perth, Adelaide and Cambridge, the new Department of Civil Aviation made a strong contribution to the provision of aerodromes in the Australian outback. The small isolated communities at outstations and small towns in the Northern Territory, Queensland and Western Australia did not have the ability to construct and maintain aerodromes and could not therefore benefit from air transport. During the dry season, supply trucks provided the needs of these communities, however in the northern tropical monsoon wet season, the roads were impassable. Numerous attempts to use packhorses could not cope with the demand because they too experienced extreme difficulties in getting through from departure point to their destination. The Department of Civil Aviation formed small Mobile Maintenance Units (MMUs) mostly using war surplus civil engineering vehicles and plants. DCA engineering and technical staff conducted the field surveys, located and tested suitable materials and also designed the aerodrome for the various communities. The crews of the MMUs were another breed of Australian who were never recognised for their hard work, endurance and loyalty in the face of extremely harsh living conditions and isolation from their families. They lived in the bush extremes from swamps to arid desert areas often cutting through virgin country for lengthy distances and frequently showing great Australian ingenuity and improvisation. Many stories have been told of the rugged characters who performed all of these tasks. There were virtually no roads in these areas and the only tracks were very rough indeed. Each year at the end of the wet season, DCA's Airport Inspectors set off in four-wheel drive vehicles from their bases in Northern Australia to inspect the new and other remote aerodromes and arrange for any maintenance work needed to effect repairs for the safe continuation of air transport. Often they were also away from home for up to 10 weeks at a time. Radio communications kept them in contact with civilisation and many of the Inspectors were obliged to use the older style of pedal powered radios developed by Alfred Traeger at the inception of the Royal Flying Doctor Service. The personalities of the era were many and their camaraderie along with their typically Australian brand of outback humour is sadly disappearing. The personalities of the old Department of Civil Aviation were wonderful, each with their own eccentricities but unswervingly loyal to their employer and the industry they served. One of these personalities worked in the Mobile Maintenance Unit and was known to everyone around the country as "Whiskers" I suspect his name was Dave but his nick-name came from the obvious fact that he did not shave and unruly whiskers trailed out around each side of his neck especially as he drove his self propelled pneumatic-tyred roller on the various aerodromes. Most people in the MMU disliked driving the roller because it was dull, repetitive and unskilled. But Whiskers seemed to thrive on it, driving up and back each day exactly as the works boss had instructed. Unlike today's uncaring union orientated worker, Whiskers not only worked long hours but also looked after the roller lavishing great care on its maintenance and appearance. He always had a few drinks every night and became a little merry and occasionally took a bath, but for the most part his appearance was reminiscent of a badly soiled unmade bed and even the bush townsfolk were in awe of his appearance. However, on one occasion, the humdrum existence of Whiskers received a shock. A mate down south in Adelaide was getting married and asked Whiskers to be his best man. This transformed Whiskers into a man of importance and action and he warmly accepted the honour that had been bestowed on him. Like his unswerving attention to his job rolling on the aerodrome he would also do things properly for his mate.

Aerodrome works in progress at Port Hedland in 1972 with contractors supplying materials for the DCA MMU. MMU roller bottom right of picture. Author

Doug Whitfield

When the day came for Whiskers to depart for the city, his fellow workers in the MMU became a little uneasy as the departure time came closer but Whiskers could still be seen working away carrying out his duties on the aerodrome. When the Fokker FK27 aircraft arrived overhead ready for landing, it seemed quite clear that Whiskers was not going to have time to clean-up before his scheduled departure. It was only as the aircraft turned onto its final approach to land that Whiskers dutifully vacated the aerodrome and trundled back to the air terminal shed. Everyone stood aghast as the typically uncommunicative Whiskers replenished the fuel in his machine, checked the oil and finally walked empty-handed at an unhurried pace up to the Fokker. He was last aboard and the door banged shut. The oil bespattered dusty plant operator was on his way South without changing his clothes or engaging in even the most rudimentary toiletries. News soon filtered back to the township that Whiskers had been a model best Man for his mate's wedding. In fact his performance was reported to have been exemplary and his entire conduct attentive and supportive. The people of the township marvelled at the news and commented that the wedding must have been one of the most odd events on record with the bedraggled Whiskers in the wedding party. Can they have been wrong? The day came for Whiskers to return. The aircraft was on time and the passengers streamed out as soon as the doors opened. The last passenger to step down from the aircraft was vaguely familiar to the more discerning viewers. He was short, immaculately groomed and attired in a very expensive suit with equally costly hat, tie, shirt, shoes and glistening cuff links. Most thought that this person was a millionaire or at least a wealthy man about town! Suddenly, someone shouted aloud, "Strewth that's bloody Whiskers!" It was indeed Whiskers. He walked at an unhurried pace past the astonished audience. He looked neither to the left nor the right and continued on until he was alongside his beloved self-propelled pneumatic-tyred roller. On reaching this object of his affections, he paused for a moment and touched its side tenderly with his fingertips. Then in full view of the now completely silent crowd, he ascended to the controls, started the engine, engaged the appropriate gear and drove onto the aerodrome to get some rolling done before the Fokker taxied out for departure. At that time Whiskers would certainly have been the best-dressed roller operator in the world to say nothing of the MMU or even the Australian outback. As the days passed, the fine clothes were ruined, the facial hair re-established and Whiskers resumed his old routine of life and his traditional very dishevelled and soiled appearance. In a few weeks as was its charter, the MMU moved northward with all of its glorious personalities and continued its very constructive role in the development of the most remote areas of the continent. Of course Whiskers could always be seen working with the team wherever they were.

The Airport Inspectors were always highly respected by the outback communities and often went out of their way to help the locals with personal problems. One story about such help was related by the late Jack Davies who was Regional Assistant Director (Airports) NSW Region when he retired in 1980. The story concerned an Airports Inspector staying overnight at a hotel in a remote township. He was surprised when the publican gave him a tobacco tin commenting that "Charlie asked me to get you to put this in the bank for him when you get back to Darwin." The Inspector opened the tin and found it was full of five-dollar bank notes. The Inspector was understandably concerned and wondered aloud why Charlie would entrust such a large sum of money to a complete stranger. The Publican just stared at him and said, "Well you're the Civil Aviation bloke aren't you?" By all accounts Charlie had been fossicking for gold in the area for a long time and did not have time to go to Darwin and figured the DCA bloke was trustworthy enough to do the job for him since he was going that way anyway!

Douglas DC3 aircraft represented the mainstay of heavy commercial aviation in Australia until 1958 when the increasing costs of operating such old ex-wartime aircraft gave way to the new breed of turboprop regional airliners. The Fokker Friendship FK27 aircraft was double the weight of the DC3 and operated with very much higher tyre pressures that demanded a much higher standard of runway even though they could be reasonably operated on unsealed surfaces. However, as the frequency of the operation increased there was a need to slow the damage to rural airstrips by getting them sealed with bituminous substances. Many of the aerodromes had sealed the runway ends to save money but not the centre part since the aircraft landing could be done on the sealed section. Similarly, erosion from propeller-blast could be minimised during the initial take-off roll on a sealed surface.

In the late 1950s, jets began to enter the competition for air routes with appearance of the De Havilland Comet, Boeing 707, Douglas DC8, French Caravelle and Lockheed Electras. Australian capital city airports therefore required substantial upgrading projects to cater for these new aircraft types. At the end of 1979, Qantas had disposed of its Boeing 707 aircraft and had replaced them with B747 Jumbo aircraft. Ansett Airlines and Trans Australia Airlines (TAA) each operated eleven Boeing 727 and twelve Douglas DC9 jet aircraft. At the same time MacRobertson Miller Airlines (MMA) were operating five Fokker FK28 jet aircraft in Western Australia and the Northern Territory. A total of forty two Fokker FK27 aircraft also remained in service in support of the smaller regional centres notably in eastern states, South Australia serving Mount Gambier, the Eyre Peninsular and the track up to Broken Hill, Leigh Creek, Alice Springs, Tennant Creek, Katherine/Tindal and Darwin. Other regular air-routes included Gove to Mount Isa and Cairns to Gove.

Joint-user facilities are located at Williamtown NSW, Darwin NT, Tindal NT and Townsville North Queensland. These unique facilities are shared by the Royal Australian Air Force and civil air terminal operators. As to be expected, military technical specifications for aerodromes differ from their civil counterparts. Civil liabilities impose stringent systems safety requirements on facilities used by the travelling public in order to protect both the interests of the carriers and operators as well as the passengers.

This arrangement creates some interesting problems where civil operators have to maintain high standards of safety in order to acknowledge insurance underwriter's liability and risk models for maximising passenger safety. The only Australian capital city that employs a joint-user military and civil terminal is Darwin and this creates some unique problems for its commercial development. The Defence Department have no interest in commercial business and safety arrangements and as owners of most of the aerodrome, they cannot operate efficiently because of the huge technical staff turnover caused by military postings. The complexity of Civil Aerodromes demand a permanent staff who get to know the idiosyncrasies of its working systems. Curiously, the standards for aerodrome visual markers, lighting facilities for night operations and regulations differ and the insurance under-writers for civil operations have to be sure that the safety of civil operations is not reduced. The Civil Aviation Regulations impose firm requirements but in reality, these are difficult to enforce when they are overpowered by the Defence Act.

The domestic and international terminal was located in a war-damaged hangar built in 1940 and converted for commercial passenger use after the war. It was damaged by the vagaries of various cyclones notably Cyclone Tracy, when it destroyed Darwin in December 1974. From 1974 the building had been so badly weakened that it was deemed unfit for use if the wind speed exceeded 130km/hr when weather forecasts indicated that such winds were possible, the terminal needed to be evacuated and shut-down on each occasion. Enough was enough and in 1984 the Federal Parliament approved the construction of a new civil aviation terminal at an estimated cost of $95 million and work started very quickly in the following month. However, in April 1985 the Minister for Aviation announced that work was to cease while the project was re-evaluated but not before some $8 million was lost through cancellations of contracts! A small air terminal was opened on the northern side of the aerodrome by the Rt. Hon Bob Collins in December 1991 and later the Federal Airports Corporation handed over the Aerodrome Licence to a private operator.

By 1992, the Federal Department of Transport and Regional Development (DoTRD) had re-assigned the responsibility for operating aerodromes that had previously been the domain of the Federal Authorities to local operators with the exception of Sydney Airport. Curiously, the size of the population outside the capital cities had not changed much when this initiative was carried out. This meant that the revenue obtainable from air operators for landing and air terminal charges was very low commensurate with the air service support for such small communities. As soon as the Airport Local Ownership Plan was implemented by DoTRD however, costs soared with the result that the air services ceased and airports became unable to pay the huge costs of running a licenced asset. The natural result was that well over a hundred airports in Australia cancelled their Airport Licence amid deteriorating standards. A well-known government controlled airport in the northwest of Western Australia once had over 200 government employees working on the aerodrome. By comparison, there are only three employees remaining trying to maintain the standards!

The Aerodrome Inspectors of today are not much different to their counterparts half a century ago. They are professional and above all practical. They have to be practical in order to keep in step with the changes in aircraft operations that prevailed over a long period of time. They are among the best risk managers in the aviation industry and can be described as the silent achievers. They receive little recognition for their dedicated work, yet to a man, they remain motivated, enthusiastic and do the best they can to help the aerodrome operators under their control to achieve safety standards. As I previously stated, with very low income from landing and other charges, this is a fairly difficult task. The Inspectors travel long distances to audit a wide variety of airports from International facilitates down to the tiniest outback airstrip that is still able to support passenger carrying aircraft operations.

The Aerodrome Inspectors usually have a good sense of humour and often need it with some of the things that happen to them. Val Augensen was a well-known Inspector who had many adventures in the 1930s and 1940s. Like many Inspectors he was also a pilot of some accomplishment. I am reminded of a story that was related by Inspector Val Augensen to Phil McCullough, a well-known Aerodrome Inspector from South Australia now resident in Western Australia. I knew Phil when he used to fly the DCA Beechcraft Bonanza aircraft up and down the track from Parafield to Darwin covering a huge number of aerodromes in both South Australia and the Northern Territory. The story was about two Aerodrome Inspectors in the 1960s, who were obliged to conduct an audit in December on an outback property located on the Finke River in the Northern Territory. December is a very hot time of the year in the Territory with temperatures not unusual at 40 to 45 degrees Celsius, and with no air conditioning available in those days, the prospect of such a trip was daunting to say the least. Even with air-conditioned aeroplanes and motor vehicles available today, the trip suffers a certain amount of tedium.

The two Department of Civil Aviation officers had been sent into this inland furnace because this job could not wait until the New Year. Sadly, they could not join in the pre-Christmas celebrations that were going on at their home base and sadly navigated their lumbering Willys Four Wheel Drive vehicle through the scorching red sandhills and dry riverbeds of the inland summer. Periodically, they needed to change and repair a flat tyre in the all-pervading "bulldust" using tools, which rapidly became too hot to handle and could only be used with wrapping from various pieces of rags. Hour after hour the Inspectors battled along the bush tracks leading to the homestead and at last in the shadows of the evening they arrived at the front gates of the desolate homestead.

The next trauma that would soon confront them was a drive over the Gibber stones to the aerodrome where they would spend the night in the open desert as two isolated outcasts, hundreds of kilometres from the nearest source of a refreshing alcoholic drink! The airstrip at the aerodrome blended into the surrounding country so well, that aviators had great difficulty in locating it from the air. However, the homestead itself was set in a neat yard on the bank of the Finke River and the Inspectors walked up the pathway to the front door of the homestead to announce their presence to the owner. They knocked at the large weather-dried door and while they waited commented to each other that even the door looked like it could do with a nice cold beer! Searing hot dry air blew as they listened to heavy footsteps approaching the door from within and as it opened they were met with the sight of a large man in his forties, dressed in khaki shirt, slacks and high-heeled stock boots. He was as drunk as the proverbial and swayed in the doorway as he surveyed the visitors. "Come in and have a drink whoever you are, enjoy yourselves, be my guest. The missus is in Adelaide and I have stockpiled all sorts of grog in vast quantities so there is plenty there to share with me. It's good stuff. No hooch!"

The two public servants watched this apparition with widened eyes and mouths agape. They suggested that a drink should be delayed until they had given the aerodrome a quick look-over but their host turned and roared, "I am expecting a visit from some DCA lunatics soon and I propose to give them the "rush" treatment. Don't suppose you damaged your eyes by seeing the fools during your trip out here did you? Come inside and do something useful like having a drink with me. Forget them. They are parasites on the tree of progress." The need of a drink combined with the gathering darkness of night caused the pair to wane and they proceeded to try and make their host happy by joining him for a drink. However, the binge continued until 4am as they progressed through their host's beer and everything else including exotic top shelf liqueurs. However, a vast range of subjects were discussed although despite their alcoholic intake, they were careful not to mention the aviation taboo. They slept in chairs until sunrise and then staggered to their feet in the intense heat and white sunlight to make a scheduled radio call at 0830am to their home base. The "sked" was made and this avoided awkward questions for later on when they returned and also prevented search parties being sent out. The terribly ill pair located their host but he was in a deep sleep amongst the aroma of stale booze, cigarette ash and that peculiar smell that seems to pervade many old houses. Fortunately for the pair, the Willys Four Wheel Drive seemed to know the way to the aerodrome where a brief inspection showed that it was serviceable and so they headed out again to locate the next place on their itinerary.

But now fate took a hand. 1960 was in the middle of the merciless 1956-1965 drought in Central Australia and the entire open countryside was comprised of loose soil through which gaunt cattle tried to eke out a living. Even the Mulga scrub was dying. On their endless treks to the water bores, these unfortunate animals created vast numbers of radial paths, which completely obliterated the bush roads. The two ailing DCA officers met several of these heavily tracked areas during their painful drive. On each occasion it was necessary to swing out for a number of kilometres to less disturbed country and then proceed along an arc to intercept the various roads. As time moved on, it became evident that something must have gone seriously wrong with the basic operation of the Universe or alternatively that the crew may have taken one or more wrong roads. The sun was located in an odd position and no recognisable landmarks could be found. Roads petered out and this required heartbreaking backtracks involving getting dry-bogged in the same places on the return journey. This was not pleasant for two suffering from hang-overs from the night before. Near sunset, with the sun still located in an odd position, the pair encountered a well-used road and their spirits soared as the kilometres rolled by. At last the shadows of evening grew long and the feelings of the two officers turned to euphoria as a homestead appeared in the evening gloom. They drove up to the rear of the house and happily banged on the door.

Heavy footsteps approached the door from within and as it opened the pair could just make out a man in his forties dressed in a khaki shirt, slacks and high-heeled stock-boots. He was clearly as drunk as the proverbial and swayed in the doorway as he surveyed the visitors. "Come in and have a drink whoever you are, enjoy yourselves, be my guest. The missus is in Adelaide and I have stockpiled all sorts of grog in vast quantities so there is plenty there to share with me. It's good stuff. No hooch!" "I am expecting a visit from some DCA lunatics soon and I propose to give them the "rush" treatment. Don't suppose you damaged your eyes by seeing the fools during your trip out here did you? Come inside and do something useful like having a drink with me. Anyway forget them, they are parasites on the tree of progress." The two hapless DCA Aerodrome Inspectors stood staring in deep shock as they resumed the same chairs as the night before. Both were now very aware that they had taken a wrong turning in the bush and driven in an enormous circle around the outback to arrive back exactly where they had started! I am sure they would have marvelled at the wonders of modern navigation with GPS used just as much by travellers on the surface as well as those in the air. Such were the adventures of the DCA officers of the time whether they were MMU plant operators, Aerodrome Inspectors, Aerodrome Engineers, Flight Service Officers or Aerodrome Groundstaff.

The Winds of Change

A new era is here and I am sitting with the captain on the flight deck of an A320 Skystar jetliner taxying at Perth Airport. The Perth Airport Pilot Briefing Office has ceased operations and with it has gone the Operational Control Centre, Rescue Coordination Centre and Meteorological Briefing Office. We are on a scheduled flight to Melbourne and we will fly across three states on the four-hour journey to our destination. The crew are the new breed trying to get used to a new system. They are survivors of economic chaos in the airline industry caused by industrial action that changed the aviation world, as we knew it in the late 1980s.

The Pilot Briefing Office at Perth Airport one-week before it closed its doors forever in 1992. Behind the counter on the left are the Operational and Meteorological Briefing Officers and the desks at the right are where the pilots used to plan their flights. (Author)

Businessmen with no proper employment experience in aviation have influenced politicians and been allowed to interfere and wreck a great Australian system with false ideals and theories that will cause changes that cannot realistically be supported by such a small Australian population. Aviation systems that work in other countries will not work in Australia until the population is more than doubled, especially in areas other than the capital cities. The current failure of the regional centres to cope with the expense of privatised airports is proof enough of where the industry is heading. We have seen the concepts of user-pays, inappropriate cost-cutting measures and the imposition of false economies. The resulting abolition of Flight Service and the containment of the entire Australian Air Traffic Control system into two main centres located at Brisbane and Melbourne to replace an Australian system that was unique, world class and almost fool-proof creates a great sadness for me and many others of our time.

Flight engineers, Navigators and even Operations Staff on the ground have now been replaced by computers and microwave links and the need for an HF radio system over the country has been superceded. With the changes, Flight Service and Operational Control staff have been made redundant. Search and Rescue alerting is now activated by surveillance satellites and the huge aeronautical telecommunications networks have been replaced by facsimile machines and computer data links. The cockpit of this aircraft seems alien and strangely clinical. No longer are there chart desks and volumes of engine operational data books. The cockpit instruments have all vanished and the pilot's panels are occupied by two visual display units that show all the data necessary for flight. The new jargon refers to this as a 'glass cockpit'. The on-board systems computer will dictate to the pilot the actions required via electronic check lists and then tell him what fuel load is required, what speeds and heights he may fly, and which route he must take. If an emergency occurs, the flight computer will still dictate the terms and the pilots must obey all the commands given. The operational parameters are even transmitted automatically to the company headquarters via a radio link, so that checks can be made on how each crew are operating the aircraft. A voice that I recognise from earlier days calls from the control tower and clears us for departure. The captain executes a smooth rolling takeoff and we accelerate rapidly into the summery haze above Perth. The traditional pilot's control column has been replaced by a side-stick control situated on a panel on the side panel to the left of the Captain and the right of the First Officer. The control-column used to dominate the position between the legs of both pilots and now leaves a large area which is occupied by a 'pull-out' tray instead, on which to write in-flight data on the flight plan forms, or used simply as a table on which to have their dinner.

We pass swiftly through 3000 feet and begin a turn to the east over the Darling Scarp that rises as a natural backdrop to the Perth area. Below us is the Avon River that hosts the annual whitewater canoe races, which are run when the waters are at their peak, raging down the valley. Today it looks rather serene as it stretches away into the distance towards the tiny hamlet of Toodyay. We continue in a gentle turn to the right to intercept our outbound track and as I look over my right shoulder behind the copilot, I can see the deep ribbon of blue of the Indian Ocean. On the left-hand side of the aircraft are the towns of Northam and Beverley which are both home to large gliding clubs. On very early mornings, when the winds are calm, balloons can be seen drifting lazily across the countryside, splashing it with their colourful envelopes. Each basket can carry up to three people and they can relive the contentment experienced by the early airmen who knew nothing of the demonic complexities that represent today's modern flying. Examination of updated weather information shows that the visibility is likely to be very good along the south coast. The crew plan a diversion which is approved by air traffic control so that the passengers may benefit from the coastal panoramas along the Great Australian Bight. The route selected also provides an increased tailwind which means we will arrive in Melbourne a lot earlier than was originally planned. I have flown all over this country and strain my eyes to see the areas where I first learned to navigate under the guidance of Arthur Turner. Memories of my very first solo cross-country flight flood into my mind as Cunderdin passes under our left wing tip. On that trip I departed from Jandakot in clear conditions but an un-forecast ground fog developed in just a few minutes, concealing the whole area over which I was trying to visually navigate. All I could see was the tops of the television station masts sticking up above the blanket of fog. What a dilemma. I didn't know where I was, I couldn't see where I was going and was likely to get lost if Perth radar couldn't identify me. I remembered an amusing story I had recently heard about a pilot who was flying low in the top of some thick ground fog, so that he could enjoy the illusion of high speed as he whizzed along in the top of the fluffy veil. Ahead of the aircraft he was startled by the sight of a man who stood up in the top of the cloud waving cheerily as the aircraft zoomed past. As the pilot flew past he saw that the man was actually standing on top of a radio mast concealed by the fog blanket below! I smiled as I thought of the incident, and returned my thoughts to the fog still lurking below me. On that trip so long ago, Flight Service had radioed to tell me that the fog was expected to clear in approximately 45 minutes and asked for my intentions. My flight time to Cunderdin was also 45 minutes and I decided to continue on my flight planned heading, in the hope that it was accurate enough for me to locate Cunderdin airfield when the fog cleared. Fortunately, my decision was correct and I landed safely.

Cunderdin and its memories are well behind us now and all seems very tranquil up here in the jet as we pass 25,000 feet in the climb to our cruising level of 35,000 feet. The crew sit silently monitoring the computer information being displayed and sip cups of coffee provided by the cabin staff. Looking to my left again I can see the countryside that encompasses the towns of Southern Cross, Norseman, Kalgoorlie, Leonora and Laverton. Here many thousands of charter flights have cris-crossed the landscape, providing a variety of air charter services to the remote settlements and stations. I have flown quite a few hours here myself and am pleased to see Kalgoorlie, just visible on the horizon with the sun glinting off some of the rooftops. Fond memories of Kalgoorlie have been clouded by the closure of the old airport that I knew so well, to make way for a new airport further south of the township. I wish my old Headmaster who said I would never fly, could see me now. I would have loved to walk in on him in my pilot uniform. Never mind, guess he will never know now. The Great Australian Bight slides by under the right wing and Esperance Township disappears under the nose as we head eastward. Deep blue waters edged by white surf crash onto the pristine white-ribbon of beaches and stretch far into the east and west horizons. We fly over Ceduna airport and its seaport of Thevenard and pass from Western Australia into South Australia. The Eyre Peninsula looms into view with Port Lincoln on our right and Whyalla on our left. A thousand voices from the past flood my memory as I remember the days spent in Flight Service in this area where we monitored air services flying all over the peninsulas of the Spencer Gulf. Our Skystar slips past Adelaide and the Victorian Wimmera district shimmers in the heat of the day and very soon the Grampian Mountains appear on the horizon. We descend into Melbourne and the Tullamarine Jetport can be easily seen in the distance. As we approach the city, I can see Mount Dandenong rising 2066 feet above sea level to the east. On a site near this peak stands a memorial to those who perished on the DC2 'Kyeema' on 25 October 1938. The service that rose as a result of that crash has now disappeared and with it has gone the skills, judgement and knowledge that a computer controlled service will never know. Computers cannot tell you a joke on a rainy night to set you on your way. They cannot provide you with a cup of coffee when the weather is freezing cold before you depart and they cannot replace a well-worn map or provide personalised service about weather and operational information either. The National Aeronautical Information Processing System (NAIPS) has only two briefing offices located in Brisbane and Melbourne. I still find some difficulty in understanding just why it is, that pilots embarking on a flight from Perth to Geraldton in Western Australia have to notify their flight plans to a briefing officer in an office 2000 miles away on the other side of the country! The Skystar crew still remember the services that used to be provided and grumble that they would still rather visit a briefing office in person instead of having to interrogate a computer. We land safely and I smile faintly, thanking them for having me aboard. Outside I take a taxi to Essendon airport where I first arrived twenty-two years ago and embarked on my aeronautical career in Flight Service, commercial flying, air traffic control and airport management. The bustling airport I remembered from then is now very quiet and seems to be used only by small commuter aircraft. The control tower is still operating and the aerodrome beacon on its roof revolves mournfully, flashing its green and white light signals to guide pilots safely into its circuit area. No doubt, this airfield will soon succumb to housing developments and its history will disappear without trace. The airport facilities are in disrepair and many buildings stand vacant. The operational and meteorological (OPS/MET) building stands as a monument to a forgotten era.

The Siemens three-row teleprinter/telex machine featuring a torn-tape message receiver and transmitter facility. Used by most DCA Air Traffic Control, Flight Service and Communications Units in the 1970s and early 1980s. (Author)

Here pilots congregated, studying weather reports and notices to airmen (NOTAMs). Briefing officers checked the flight plans and submitted them for relay to all the appropriate Flight Service and Air Traffic Control units via the Aeronautical Fixed Telecommunications Network (AFTN) teleprinter service.

As I approach the building, I find the doors are barricaded and padlocked. Paintwork on the once neatly kept facade, peels off the window frames leaving the wood beneath it bare. Inside, the once highly polished black and white tiled floors are dulled forever.

The DCA OPS/MET Building at Essendon Airport now standing derelict with a thousand memories in 1992. In its day the highly maintained building's Art-Deco façade included flying the Civil Ensign and the Australian National flag from the flag-masts and imbued a sense of pride in those who worked there. (Author)

Trans Australia Airlines (TAA) Lockheed Electra turbo-prop airliner "John Eyre" on the tarmac at Essendon Airport in January 1970 (Author)

The staff have long since left and only the footsteps of their ghosts remain. Gone too are the mighty four engine turboprop Lockheed Electras and the throaty roar of the radial engines on the Douglas DC3 inter-city services. This lament is carried on around Australia and the units have closed down their operation and an excellent service forever. As I walk slowly along, reliving this rather grand history, I ponder the DCA College motto "In pursuit of excellence." It reflects the labours of those who excelled by toiling in arduous conditions often in remote locations, dedicated to their service in all circumstances. A rather cynical, though sad message transmitted by the Alice Springs Flight Service Unit on its last day of operation on 10 December 1991 reflected the general feeling. Those who received the message, knew that it would soon be their turn to become redundant and that the aeromobile HF frequencies would be silenced forever. I stare around me, determined that some record should be made of these times, so that our descendants know of the part it played in aviation history. Bureaucracies are not interested in preserving history and those of us who took part in the evolution and demise of the old services must try and preserve their stories. The air services that daily lined up into the wind, ready for takeoff, no longer exist, victims to the ravages of time. Now as I stand lost in my thoughts, the sky darkens and I am warned of an approaching storm. Momentarily, I reflect that the wind that blew in those early days is still the same wind blowing now and I salute my Flight Service and Operational Control and all past DCA comrades for their service in pursuit of excellence.

```
102341
FF APPPZGZX
102340 ASASYOYX
ALICE SPRINGS AIR/GROUND WILL BE CLOSING AT 0730UTC AFTER 52 YEARS
BOTH MAN AND BOY. THE MAJORITY OF THE STAFF WILL BE LEAVING FOR AD
AND ML. THE BO WILL REMAIN OPEN, MANNED BY 4 DEDICATED PRE-
ANNUANTS OF MATURE AGE AND JUDGEMENT
A SMALL SOIREE WILL BE HELD IN THE NEW DLRS22 MILLION TERMINAL TO
BID FAREWELL.
MESSAGES OF CONDOLENCE OR CONGRATULATION WILL BE RECEIVED AT
THE ABOVE ADDRESS BY 0700UTC.
NO FLOWERS PLEASE.

NNNN
```

Glossary

ADF	Used on aircraft in conjunction with a ground NDB
Aeradio	Aeronautical ground radio unit
AFIZ	Aerodrome Flight Information Zone (Usually that area around an airport within a five nautical mile radius up to a height of 5000 feet)
AFTN	Aeronautical Fixed Telecommunications Network (Teleprinter message relay system)
AIREP	Air Position Report
ATC	Air Traffic Control
AWA	Amalgamated Wireless Australia (A commercial radio builder)
BPA	Bush Pilot Airways (Cairns) (Later became Air Queensland & taken over by Australian Airlines)
CFI	Chief Flying Instructor
CW	Continuous Wave (Used for Morse signals)
DCA	Department of Civil Aviation (Replaced by the Department of Transport)
DDS & NA	Department of District Services and Native Affairs
DME	Distance Measuring Equipment (Gives the pilot a read out of distance to the beacon in nautical miles, (max range approx 125 miles)
FIA	Flight Information Area
FSC	Flight Service Centre (Responsible for the FSU's in a Flight Information Region)
FSO	Flight Service Officer
FSU	Flight Service Unit (Replaced the Aeradio Unit and was responsible for a designated geographical Flight Information Area)

GMT	Greenwich Mean Time (Time based on the zero meridian through Greenwich in Britain. Now superceded by UTC which is the Universal Time Constant which is based on the same meridian)
GPO	General Post Office
ICAO	International Civil Aviation Organisation
ITU	International Telegraph Union
HF	High Frequency (3 MHz - 30 MHz)
KBAC	Kalgoorlie-Boulder Aero Clu
KHz	Kilohertz (1000 cycles per second)
Knot	Unit of one nautical mile per hour (1 knot = 6088feet)
LMT	Local Mean Time
LSALT	Lowest Safe Altitude
MF	Medium Frequency (300KHz-3MHz)
MHz	Megahertz (1,000,000 cycles per second)
NDB	Non Directional Beacon (Used in conjunction with an ADF using the MF band)
NOTAM	Notice(s) to Airmen
OCC	Operational Control Centre
OIC	Officer In Charge
OTC	Overseas Telecommunications Corporation (Taken over by Telstra)
PFIB	Pre-Flight Information Bulletin (Notams arranged in groupings for certain areas or air-routes)
PMG	Post Master Generals Department
QANTAS	Queensland and Northern Territory Aerial Service
RAAF	Royal Australian Air Force
RAF	Royal Air Force (British)
RCC	Rescue Coordination Centre
RFDS	Royal Flying Doctor Service
RNZAF	Royal New Zealand Air Force

RTT	Radio teletype (Teleprinter message relay service using radio links instead of landlines)
SAATAS	South Australian and Territory Aerial Services
SAR	Search and Rescue
SARTIME	Search and Rescue alerting time
SARWATCH	Search and Rescue surveillance of an aircraft from taxying at departure point until landed at the destination. Aircraft using this system were required to communicate with a ground unit from any point from the commencement of the flight
SEA3	South East Asian HF network number 3.
SELCAL	Selective Calling Device (HF Radio Link with electronic alerting functions)
SOC	Senior Operations Controller in the OCC
SSB	Single Side Band radio system
Static	Radio slang referring to radio interference which is badly affecting radio signals received
TAA	Trans Australia Airlines (Later became Australian Airlines and was taken over by Qantas Airways)
UHF	Ultra High Frequency (300MHz - 3GHz)
USAF	United States Air Force
VHF	Very High Frequency (30MHz - 300 MHz)
VOR	VHF Omni Range (Ground navigation beacon giving the pilot information on which radial through 360 degrees the aircraft is situated and whether it is flying 'TO' or 'FROM' the beacon)
VSB	VHF Survival Beacon
WAC	World Aeronautical Chart (Navigation chart series on a scale of 1:1000,000)

Doug Whitfield

IN MEMORIUM OF THOSE WHO I KNEW WELL AND FLEW WITH WHOSE WINGS ARE FOREVER FOLDED

✝

Nigel Halsey

Arthur Turner

Jimmy Davidson

Steve Carter

Graham Brooks

Trevor Brougham

Roger Connellan

E J (Eddie) Connellan

H (Ossie) Watts

Ted Smith

Ross Fox

Gerry Ferguson

Arnold Fox

Guidio Zuccolli

Raelene Whiting

Ron ('Son') Beamont

The Flight Service and Communications Officers with whom I worked and remember
1969-1976

DARWIN
Eric Bergholtz
Jan Binnekamp
Case Schoolmeister
Bert Thrupp
Bill Orr
Syd Sneyd
Dan Dowling
John Cooper
Peter Hendrix
Tom Cooper
Roy Shepherd
Ray Shephard
Ray Thompson
Doug Whitfield
Ron Dunkin
Doug Morrison
Kev Ruwoldt
George Strawn
Gary Presswell
Gerry Wareham
Dennis Prider
Les McCourt
John Meaker
Brian Wise
John Brady

ALICE SPRINGS
Laurie Tomlinson
Tom Dodderidge
Frank Hind
Alan Miers
Dick Charleston
John Jenkins
Gordon Hemming
John Scougall
Trevor Ward
Jim Ware
Bob Watson

Kev McGee
Jack Ward

ADELAIDE
Allan Mathews
Ray Kiely
Case Vleugel
Clarrie Castle
Don Frey
Stan Mayne
Peter Bull
Austin Noblett
Les Way
John Border
Kev Brannigan
Alex Border
Terry Cout

KALGOORLIE
Ken Davies
Clive Cassin
Graham Cooney
Kevin Walters
Tim Slade
Jim Mullins
Glenys Richards

GOVE
Adrian Shee
Ron Agutter
John Sullivan

KATHERINE
Don Middlemiss
Don Hodder
Jim Brady
Leo Sheridan
Terry Hind

WHYALLA
Howard Abbott
Brian Scherr
Laurie Byers